Sparks of Genius

IEEE Press
445 Hoes Lane, P.O. Box 1331
Piscataway, NJ 08855-1331

1993 Editorial Board
William Perkins, *Editor in Chief*

R. S. Blicq	R. F. Hoyt	J. M. F. Moura
M. Eden	J. D. Irwin	I. Peden
D. M. Etter	S. V. Kartalopoulos	L. Shaw
J. J. Farrell III	P. Laplante	M. Simaan
L. E. Frenzel	M. Lightner	D. J. Wells
G. F. Hoffnagle	E. K. Miller	

Dudley R. Kay, *Director of Book Publishing*
Carrie Briggs, *Administrative Assistant*
Valerie Zaborski, *Production Editor*
Cover Design:
Tisa Lynn Lerner / Lerner Communications

The IEEE Center for History of Electrical Engineering is grateful to the organizations and individuals listed below who provide generous support to the center in the form of operating, endowment, and project funding.

Founding Partners: IEEE
Rutgers University
IBM Corporation
IEEE Foundation - General Fund
IEEE Foundation - Life Member Fund
Andrew W. Mellon Foundation
Alfred P. Sloan Foundation

Senior Partners: National Science Foundation

Partners: AT&T Foundation

Colleagues: Electron Devices Society
Microwave Theory and Techniques Society

Associates: Antennas and Propagation Society
Electro-Mechanics Company
Environmental Research Institute of Michigan
KBR Foundation
Sematech
Takashi Sugiyama

We are also grateful to the thousands of individuals who make annual contributions to our Friends Fund.

Sparks of Genius

Portraits
of Electrical Engineering
Excellence

Frederik Nebeker

with contributions by

William Aspray, Andrew Goldstein, and David L. Morton, Jr.

IEEE Center for the History of Electrical Engineering

IEEE PRESS

The Institute of Electrical and Electronics Engineers, Inc., New York

This book may be purchased at a discount from the publisher when ordered in bulk quantities. For more information contact:

IEEE PRESS Marketing
Attn: Special Sales
P.O. Box 1331
445 Hoes Lane
Piscataway, NJ 08855-1331
(908) 981-8062 (FAX)

Printed in the United States of America

10 9 8 7 6 5 4 3 2 1

ISBN   0-7803-1033-0
IEEE Order Number: PC03822

Library of Congress Cataloging-in-Publication Data

Nebeker, Frederik.
    Sparks of genius : portraits of electrical engineering excellence / by Frederik Nebeker.
      p. cm.
    Includes bibliographical references.
    Contents: Ernst Weber, bridger of cultures—Working to establish the discipline of biomedical engineering, the career of Herman P. Schwan—From radar bombing systems to the maser, the engineering experience of Charles Townes—Gordon Teal, finding the right material / by Andrew Goldstein—Harold Alden Wheeler, a lifetime of applied electronics—Calculating power, the career of Edwin L. Harder / by William Aspray—New applications of the computer, Thelma Estrin and biomedical engineering—The magic of your dial, Amos Joel and the advent of electronic telephone switching / by David L. Morton, Jr.
    ISBN 0-7803-1033-0
    1. Electric engineers—Biography.
TK139.N42   1993
621.3'092'2—dc20
  [B]
                                                  93-15916
                                                    CIP

CONTENTS

FOREWORD

This book is a product of the Center for the History of Electrical Engineering. Founded in 1980 by the Institute of Electrical and Electronics Engineers, Inc. (IEEE) and carried out as a joint venture with Rutgers—The State University of New Jersey since 1990, the center promotes the study and understanding of electrical science and technology through historical research, education, public outreach, and archival collection.

Much of the material for this volume resulted from the IEEE Oral History Program, which is made possible by the generous support of the IEEE Foundation Life Member Fund. The objective of this program is to preserve one of the most valuable assets of the IEEE, the memories and understanding of its senior members. Under this program comprehensive interviews have been conducted with some of the leading electrical engineers; edited and indexed transcripts of these interviews have been prepared.

It is our hope and expectation that this book will advance objectives that are dear to us at the center: to encourage careers in engineering; to provide role models for young men and women; to increase public understanding of what engineers do and how their work affects daily life; to illuminate the scientific, business, and social dimensions of engineering; and to document some of the great achievements of electrical engineers.

William Aspray
Director, IEEE—Rutgers Center for the
History of Electrical Engineering

INTRODUCTION

This book tells the stories of eight engineers. They were, with one exception, born before the end of World War I, and so into a world very different from ours today. Though electrical engineering was a well-established profession—the American Institute of Electrical Engineers dates back to 1884— it was not the multifarious profession it has since become: then the generation and distribution of electric power was the main concern of most of its members.

The first two decades of the century were years of rapid expansion for electric power. The share of mechanical power in US industry that electricity provided grew from 3.6 percent in 1899 to 52.5 percent in 1920. This was the great age of the electric trolley: beginning in 1887 trolleys contributed vitally to the growth of cities, and interurban lines proliferated—reaching a total mileage of 15,000 in the United States. Electric lighting had changed the urban landscape, with many cities having their own "White Ways" by 1918. And it was becoming possible to reach more and more people by telephone; in 1915, for example, service between New York and San Francisco commenced.

Yet it was only in cities that electricity in the home was at all common, and most homes that had electricity used it only for lighting. Though many offices and homes had a telephone, it almost certainly had no dial, and almost everywhere an operator connected each call by hand. Public address systems and electric traffic lights had just begun to appear. Regular radio broadcasting did not exist, nor did sound movies, tape recorders, television, or VCRs. Nor were there electric typewriters, electronic computers, radar, photocopiers, lasers, or satellite communications.

The world has indeed changed enormously since World War I, and electrical technology—including electronics and computer technology—has been a principle force in this transformation. This much is widely appreciated. But how many people, if asked to name outstanding electrical engineers, could list more than a few inventors, such as Morse, Bell, and Edison, who predate this transformation? From the outside—even from other technical areas within electrical engineering—all that is usually seen is the technology, and its advance is attributed more often to science or to the marketplace or to government demand than to the ingenuity of engineers.

It is characteristic of our culture that literary, artistic, or scientific work bears the name of the writer, artist, or scientist, while engineering work is usually anonymous. There are no doubt many reasons for this, such as that most people have little understanding of technology, that the collaborative nature of modern engineering means that achievements are seldom attributed to individuals, and that engineering is often regarded as routine problem-solving (and even the obviously innovative engineering is often thought of as merely applying the discoveries of scientists).

Although the view of engineering as routine is perhaps the most important of these factors, it is, as any engineer knows, a misperception. Engineering draws on a large body of knowledge, some of it embodied in rules. But these rules seldom determine, in any large measure, the solutions that are arrived at. The individual engineer—his or her knowledge, experience, talents, insight, and inspiration—matters in the way each task is accomplished. This much is true of all engineers. Beyond this, there are many especially creative engineers who put their individual stamp on engineering in a larger way by helping to determine the direction of technological advance. This they may do by augmenting the stock of available technological means, by recognizing that existing means can be applied in a new way, or by directing the work of a large number of engineers.

It is the purpose of this book to spotlight the individual engineer. This is important for understanding how, in general, technological advance comes about and how, more particularly, certain technologies of the late twentieth century have attained their present form. What's more, the story of the individual engineer is very often—it's certainly the case with the eight subjects of this book—a fascinating one, being a part of the momentous political, military, and scientific developments of our rather tumultuous century. We have selected eight direction-setting engineers, not with the intention of choosing the most important engineers living today, but as a sampling of distinguished electrical engineers from a wide range of technical areas.

Ernst Weber, born in Vienna in 1901, earned doctorates in both physics and electrical engineering and worked as an engineer for Siemens-Schuckert before moving to the United States in 1930. For the next 39 years he served Polytechnic Institute of Brooklyn (later Polytechnic University) as researcher in microwave techniques, teacher, and administrator. He was president of Polytechnic from 1957 to 1969, and in 1963 he became the first president of the newly formed Institute of Electrical and Electronics Engineers.

Another European immigrant engineer is Herman Schwan, who was born in Aachen, Germany in 1915. Hardships and hindrances associated with the Nazi regime delayed his education and led him to a position at the Institute for the Physical Foundations of Medicine at the University of Frankfurt, which was one of the very first institutions devoted to research in what today is called biophysics and biomedical engineering. Shortly after the war he emigrated to the United States, where his research, teaching, and professional activities helped establish the discipline of biomedical engineering.

During the war Schwan became involved with the new technology of radar, and the same happened in the United States to Charles Townes. Born in South Carolina in 1915, Townes earned a Ph.D. in physics before beginning work at Bell Telephone Laboratories in 1939. In early 1941 he was asked to help design, build, and test radar bombing systems. This work led to his development of the new field of microwave spectroscopy and later to the invention of the maser and laser. These devices opened the new realm of quantum electronics, which has already resulted in hundreds of important practical applications.

Another Bell Labs scientist who had a profound influence on the development of electronics was Gordon Teal. Teal, who was born in Texas in 1907 and trained as a chemist, worked at Bell Labs and later at Texas Instruments. He played a crucial role in the emergence of solid-state electronics by recognizing the value of fabricating components from single crystals, and he was the first to develop techniques for producing monocrystalline germanium and silicon. The story of his career shows how materials science became an important part of the electronics industry.

The main source of the electronics industry was, of course, the radio industry. The career of Harold Wheeler shows well the transition from radio engineering to the wide-ranging electronics engineering of today. Born in South Dakota in 1903, Wheeler made numerous contributions—theoretical, instrumental, and practical—to the development of radio, television, and radar. His most famous invention is a circuit that achieves automatic volume control (now called automatic gain control) that became, and remains today, a standard feature of AM radios. During and after World War II the design of antennas and transmission lines became specialties of his.

Transmission lines were also a concern of Edwin Harder's. Harder was born in Buffalo, New York in 1905 and worked his entire career at

Westinghouse Electric Company. There his calculational approach to power engineering led to many patents and practical devices. Just after World War II he directed the design, development, and application of a general-purpose analog computer, the Anacom, which remained in service at Westinghouse until 1991.

While the power engineer Edwin Harder came to be involved in analog computing, the biomedical engineer Thelma Estrin came to be involved in digital computing. Estrin, who was born in New York City in 1924, entered engineering because of the urgent need for workers in the defense industries during World War II. She went on to receive a Ph.D. in electrical engineering and became a pioneer in two respects, first in the application of computers to biomedical research and health-care delivery, and second in providing a role model for young women considering a career in engineering.

The computer has become a universal tool, now vital to many branches of engineering. Another engineer whose career illustrates this is Amos Joel, born in Philadelphia in 1918. His fascination with telephone switching began in his boyhood and is just as strong today. During the years in between, most of them spent at Bell Labs, he made countless contributions to that field. He played a central role in the introduction of electronics and computing technology into telephone switching in the decades since World War II.

These eight engineers were certainly direction-setting. They augmented the stock of available technological means in, for example, Teal's single-crystal transistors, Harder's analog computers, or Townes's maser. They adapted and extended existing means to open new areas of engineering, as Joel's introduction of electronics into switching systems, Schwan's development of electrical instrumentation to understand biology, and Estrin's use of computers in biomedical research and health care delivery. And they directed the work of other engineers, as Weber at Polytechnic's Microwave Research Institute and Wheeler at Hazeltine Corporation and Wheeler Laboratories.   .

The stories of these engineers cast light on the emergence of new branches of electrical engineering—microwave engineering, television engineering, analog computing, solid-state electronics, quantum electronics, and biomedical engineering—as well as the continued advance of older branches—radio engineering, power engineering, and telephone switching. Their stories show connections of electrical engineering with other disciplines: mathematics (Harder), physics (Weber, Wheeler, Townes, Schwan), chemistry (Teal), and biology (Schwan, Estrin). Their stories show the impact of World War II on the engineers of their generation and the way in which resourceful people managed to profit from vicissitude. Most important, their stories show that in engineering, as in science and the arts, individuals do matter.

CHAPTER 1

ERNST WEBER

Bridger of Cultures
Ernst Weber
as Researcher, Educator, and Statesman

Figure 1. Ernst Weber, first president of the IEEE, president emeritus of Polytechnic University, National Medal of Science winner, and current resident of Columbus, North Carolina.

> Young as I was, I hadn't yet understood at that time that life is the only school that counts, that its students are tested every hour of every day, and that providence calls on them to test their mettle. The limited opportunities at home, the desire to increase my knowledge, and an innate urge for action had made me cross the ocean and come to the US.

These lines come from the German author of travel and adventure stories, Karl May (1842–1912), one of the all-time best-selling authors of fiction in the world.[1] Though a popular writer, Karl May has won the esteem of many literary critics, and his books were an acknowledged influence on many German writers, including Robert Müller and Günter de Bruyn.[2] Among May's most popular novels were those set in North America, including *Winnetou* and *Der Schatz im Silbersee* (see Figure 2). These novels were full of accurate ethnographic and geographic detail, and they gave many European youngsters a strong desire to visit the New World. In 1912 one such youngster was Ernst Weber, a schoolboy in Vienna.[3] He recalls, "I practically ate up that literature; I had a great many of Karl May's works.... Already as a young boy I wanted to get to America."[4]

Figure 2. The cover of volume one of a 1951 reprinting of Karl May's Winnetou
(Karl May Verlag, Bamberg). The novel first appeared in 1893.

Viennese Youth

Weber was born 6 September 1901 in Vienna, sixth largest city in the world
and capital of the Austro-Hungarian Empire. His father, an employee of
the Austrian railway system, encouraged his interest in things mechanical,
giving him a construction set (trademarked *Steinbaukasten*) that came to
be much used (see Figure 3). The boy was just turning thirteen when war
broke out. The war—and, especially, the economic blockade of Austria that
the Allies did not lift until late March 1919— caused severe hardship to the
Viennese, and many starved to death. Weber, who helped his parents care
for four younger sisters, remembers scouring the Vienna woods for anything edible.

After the war the family moved to a new apartment, one equipped with
electricity. Weber recalls, "...the amazing thing, you push a button and the
light comes on. One cannot describe the feeling."[5] Electricity seemed just
short of magic and Weber made a hobby of building electrical gadgets,
including a crystal radio.

But Weber had other interests. Like many Viennese at that time, he felt a *Drang zum Kulturellen* because the city, even after the demise of the empire, continued to be a world cultural capital. Franz Lehár, Richard Strauss, and Arnold Schönberg carried on in the traditions of Johann Strauss, Johannes Brahms, and Gustav Mahler. Vienna theater was at a

Figure 3. Ernst Weber with two of his sisters, Josephine (*left*) and Hilda circa 1910.

high point, presenting a great variety of plays, including those of native sons Franz Grillparzer and Hugo von Hofmannsthal. There were renowned artists and architects including Gustav Klimt, Oskar Kokoschka, Adolf Loos, and Joseph Hoffman; novelists such as Arthur Schnitzler and Stefan Zweig; and philosophers such as Ludwig Wittgenstein and the members of the Vienna Circle, whose logical positivism dominated academic philosophy for much of the century.

So it was easy for Weber to acquire a taste, which he has retained to the present, for music (especially Wagnerian opera), literature (above all, Goethe), and philosophy (primarily Schopenhauer and Nietzsche). The youth was also drawn to religion. Around 1920 he converted from the Catholic faith of his family to the Helvetic Confession, an undogmatic church whose roots go back to the sixteenth-century reformer Ulrich Zwingli.

Weber was an excellent student, helped by natural talents and remarkable industriousness. He learned languages easily, which later stood him in good stead as an immigrant to the United States and as an international ambassador for his professional organization. He taught himself shorthand to use in note-taking, and after graduating from high school he pursued a dual education, taking classes at both the humanistic university (*Universität Wien*) and the technical university (*Technische Hochschule Wien*, now *Technische Universität Wien*) and often running along the Ringstrasse the two or three miles separating the universities.

The cultural attractions and the variety of courses notwithstanding, Weber was not in doubt about his career choice. More than his hobby of electrical experimentation, it was the beauty and power of Maxwell's theory that made him decide on electrical engineering. In describing his university years, he said, "I loved Maxwell's theory! From the very beginning it was to me a marvel that you could, with a few mathematical symbols, describe what goes on."[6] What impressed Weber then and throughout his career was the usefulness of the equations, which, although expressed abstractly and mathematically, have "a tremendous power of direct physical interpretation."[7] In 1949 Weber wrote an introduction to a reprinting of Oliver Heaviside's 3-volume classic, *Electromagnetic Theory*. He wrote that a key to understanding Heaviside is his great admiration for Maxwell.[8] The same could be said of Weber.

At the technical university he chose the recently instituted program in electrical engineering, and in 1924 he received the engineering degree, thus becoming a *Diplom Ingenieur*. Though he then began full-time work, he continued his education at both universities. At the humanistic university he worked with the well-known physicist Felix Ehrenhaft and in 1926 completed a Ph.D. with a dissertation, based on Maxwell's theory, of the color produced by the diffraction of light by submicroscopic particles. In the following year he earned a Sc.D. at the technical university with a theoretical study of electric and magnetic field distributions.

WEBER: *. . . Since you could observe ultramicroscopic particles only by color specks, one needed to relate the color to particle size. The theory was developed by Mie. It's a classical theory of diffraction of light by submicroscopic particles. I used the theory, but I needed to explain color relating to size. From the electromagnetic field, Maxwell's theory, I could deduce the size and then, in a color triangle, relate the color that appears. And it happened that the color triangle gave a spiral for size and color relations. That explained that there were two [size] values for the same color.*

NEBEKER: *This work was entirely theoretical?*

WEBER: *Yes. ... Another one of the same group had the assignment to carry out the experimentation. So we had additional evidence. ...*

NEBEKER: *But you had no way, directly, to measure the size of these ultra-microscopic particles?*

WEBER: *No.*

NEBEKER: *Your confidence in your derivation came from the fact that it explained the discrepancy between Ehrenhaft's experiments and Millikan's?*

WEBER: *Yes.*

NEBEKER: *... How did you find the time to do this work while working for Siemens?*

WEBER: *Well, I was young. [Laughter]*

NEBEKER: *You were working a 40-hour week or so for Siemens?*

WEBER: *Usually.*

NEBEKER: *And then working evenings on this. Was that how it worked?*

WEBER: *Yes. And of course I had to use the slide rule—there was no computer at the time.*

NEBEKER: *There was a lot of calculation involved in all this work?*

WEBER: *Sure. In fact, I had to compute some of the Bessel functions for values that were not in the tables. I had the function tables of Jahnke-Emde—they were my lifesaver![9]*

When Weber completed his engineering degree in 1924, jobs of any sort were extremely scarce in Austria. Fortunately, at his final oral examination, he so impressed one professor that the professor recommended him to the personnel manager of Siemens-Schuckert company. Siemens-Schuckert, based in Berlin, was one of the world's largest manufacturers of electrical equipment and had a major facility in Vienna. Invited to an interview, Weber was asked to look at the designs for some large generators, built for railroad electrification, whose operating efficiency was lower than expected. Weber noticed that the pole shoes were of massive iron and explained that losses could be reduced by using a laminated construction instead. This helped him to get the job of research engineer, and over the next five years he worked on the design of dynamos and electric motors. One of his assignments was to design motors for use in mines—DC motors that had to develop terrific starting moments.

In his work at Siemens, Weber constantly drew on his scientific training as he used conformal mapping and other techniques to calculate, for a variety of electrical machines, such things as mechanical forces, magnetic leakage, and the slot factor. He was fortunate in his choice of employer since the Siemens-Schuckert company was in the forefront of the effort to use electromagnetic theory to guide the design process. Werner von Siemens, the founder of the Siemens companies, believed strongly in the practical value of scientific training and research. At the Siemens-Schuckert facility in Berlin there was a development division headed by the illustrious engineer Reinhold Rüdenberg.[10]

After four years at Siemens in Vienna, Weber sought a job at the Berlin facility, which he believed would be a more challenging and stimulating environment. He was successful in this and in January, 1929, moved to Berlin, where he was design and research engineer at Siemens and adjunct professor in the department of electrical engineering at the technical university in Berlin-Charlottenburg. By the end of 1930 he had published eleven articles based on his work at Siemens; "Magnetic fields in synchronous machines under no-load conditions" and "The switching of magnetically saturated, separately excited direct-current machines" are two. They appeared in journals fostering the young science of electrical engineering: *Elektrotechnische Zeitschrift, Archiv für Elektrotechnik, Elektrotechnik und Maschinenbau*, and a research journal of the Siemens company. He also contributed to a conference that was a landmark in establishing ties between academia and industry; its proceedings, published in Berlin in 1932, were translated into English and published by MIT Press as *Theory of Functions as Applied to Engineering Problems*.[11]

United States Immigrant

Weber enjoyed the work at Siemens, but not the political atmosphere of 1930 Berlin. At this time the German economy was prostrate, and, in the view of many, war-reparation payments were holding it down. There were clashes in the streets between National Socialists (Nazis) and Communists while political moderates became fewer and fewer. The group Weber was part of at Siemens became absolutely polarized—those applauding the Nazis and those, like Weber, condemning them—and the two factions almost came to blows.

NEBEKER: *What impressions do you have of the Austrian and the German professional societies?*

WEBER: *Southern Germany, like Bavaria and regions along the Rhine, has an entirely different way of life and way of conceiving life than the northern part of Germany. The northern part is Prussian. They easily fall into regimentation. The southern people, no. The southerner is more like the Austrians, easy-going and absolutely no regimentation.*

NEBEKER: *Was that evident in these two professional societies?*

WEBER: *Yes, definitely. There were many sayings that illustrate that. A Prussian might say: "This is absolutely hopeless!" And an Austrian would say: "Yes, it's hopeless, but not impossible."*[12]

A chance acquaintance led to the offer of a one-year appointment to Polytechnic Institute of Brooklyn (later Polytechnic University) as a visiting professor in the electrical engineering department. Weber's strong desire to visit the United States—stemming from his reading of Karl May—and his unease in the German political atmosphere prompted his acceptance.[13] Granted a leave of absence from Siemens, he moved to New York City in the late summer. He recalls being shocked by the humid 90-degree heat in September and the prevalence of alcohol consumption (much greater than in Europe, and this in a country where the manufacture and sale of alcohol were unlawful).

But he found Polytechnic Institute much to his liking. It had an excellent EE program, and he was impressed with the students—"To a teacher this is the greatest compensation, that the students like what they do and want to learn"[14]—and with the New York area, which at the time was the center of radio engineering in the United States. So when, halfway through the academic year, he was offered a permanent position as research professor of electrical engineering (one of the first such professorships in the United States), he readily accepted.

The remarkable advances in engineering science in the interwar years, the increasing complexity of power networks and electrical machines, and the opening up of the realm of electronics engineering (not yet called that)— all these made employers more interested in engineers with graduate degrees.[15] Partly as a result, EE education began to make great advances, emerging from what have been called "the stagnant years" of 1900 to 1930.[16] It was in 1930 that Karl T. Compton, newly named to the presidency of MIT, began vigorously expanding graduate education at that institution. At about the same time, Frederick Terman began building an outstanding graduate EE program at Stanford University.

Weber too saw the need for graduate education in electrical engineering. Soon after his arrival at Polytechnic, he was placed in charge of graduate study in electrical engineering, for which he set high standards, both for the EE curriculum and for the faculty. His experience in industry had shown him the value of advanced training in mathematics and physics, and his experience in academia had convinced him that graduate training must be by people actively engaged in research.[17] When Weber had arrived at Polytechnic, all EE courses had to do with power engineering except for one radio course that was, as Weber says, "tolerated."[18] Weber immediately added a course on electromagnetic waves and saw to it that the usefulness of Maxwell's theory was emphasized in other courses.

Since few could afford to pursue a graduate education full time, Weber worked hard to expand an evening program that had begun at Polytechnic in 1926.[19] Most of his graduate students were simultaneously working in industry which had the advantage that they were, in general, more experienced, more motivated, and more demanding than "the day student who steps from his senior year into graduate study."[20] On the other hand, there was "the grave danger to consider as dissertation a well-written report on an assigned phase of a development program . . ." when "any good engineer would normally be expected to perform equally well just for his salary."[21] Weber always insisted that advanced degrees were more than that, and he jealously guarded the standards of achievement expected of the master's degree and doctorate. With each graduate student he visited the place of work and talked with the supervisor to ensure that the proposed project was the student's own work and that the company had no objection to the employee publishing a dissertation on his research.[22]

NEBEKER: *. . . in those years I take it the graduate program [at Polytechnic] did well?*

WEBER: *Terrifically! It grew by leaps and bounds. And it was in the middle thirties that a program of courses in engineering science, technology, and management was planned. A program to give background in electromagnetic theory principally, because in the U.S. at the time the main emphasis was on power. Communication engineering was very much subordinated. At Polytechnic there was one professor permitted to teach it. I say "permitted" because everybody else was concerned with power engineering and looked down on that playing about with little things.*

I found I learned an awful lot from my students [many of whom were working at Bell Laboratories]. Of course I knew general electrical theory; I knew magnetic circuits and electric circuits and so on well enough to realize what problems they had. Although I had started with generator design and so on, so I was also biased. I immediately became interested in propagation, partly because Maxwell's theory was an idol of mine. It was a remarkable synthesis of all electromagnetism. So along with the students I learned communication theory.

NEBEKER: *What do you mean by communication theory?*

WEBER: *The study of higher frequencies into the kilocycle range, then megacycle range. I gave courses in high-frequency analysis and so on. In fact, I laid out for Polytechnic a whole program of engineering science, technology, and management courses, with emphasis on ultrahigh frequency. That helped me in staying at Polytechnic during World War II—in spite of Radiation Laboratory trying to get everybody up to MIT—and in keeping my research group on ultrahigh frequency and, eventually, radar.*[23]

The result was that the program prospered, both in number of students and in esteem among engineers. Polytechnic awarded its first doctorate in electrical engineering in 1936. After Weber became department head in 1945, he introduced a grouping of courses for "electrophysics," which he defined as the application to electrical propagation and electrical devices of topics in theoretical physics and applied mathematics.[24] Under Weber's direction, the electrical engineering department grew steadily; by the late 1950s it comprised 38 percent of Polytechnic's total enrollment of 5500 undergraduate and graduate students in the nine fields of science and engineering at the school.[25]

Researcher of Ultrahigh Frequencies

In the 1930s Weber, interested in radio and television communication at high frequencies, investigated a part of the electromagnetic spectrum that had been little studied, the so-called ultrahigh frequencies (up to 600 MHz, which corresponds to a wavelength of 50 cm). He made studies of experimental electron tubes, some obtained from Bell Laboratories, for generating ultrahigh-frequency waves. With the outbreak of war and the need to develop effective radar systems, these and even higher frequencies (up to 10,000 MHz or 3 cm), which came to be called microwaves, suddenly assumed great importance. Weber decided to focus his efforts on techniques for accurately measuring frequencies, wavelengths, and power attenuation of microwaves.[26]

Radar—for airplane, ship, and submarine detection; for fire control; and for navigation—played a central role in World War II, but at the beginning of the war there was a need for an accurate and practical way to measure the power output of a radar transmitter and the sensitivity of a radar receiver. A radar transmitter might, because of a faulty tube or other malfunction, be emitting less power than the operator believed, in which case the range of detection would be less. So an operator might believe that he would detect any ship within fifty miles when the range of his radar was in fact thirty miles. The same effect could result from diminished sensitivity of the radar receiver.

Such problems could be discovered and corrected by frequent measurement of power output and of sensitivity, but at the outset of the war there were no simple ways to do this. Weber realized that an accurately calibrated attenuator, rugged enough to be used in the field, was needed. Working with Anthony Giordano and others at Polytechnic, Weber designed a coaxial device in which the inner conductor was a glass rod coated with a platinum–palladium alloy. In designing this device Weber used fundamental theory, especially as elaborated by himself and others in studies of the "skin effect," to calculate microwave resistance of thin films. Taking great care in the selection of materials and in the fabrication of the device, Weber's group soon succeeded in building an extremely durable instrument that was capable of precision measurement. When Weber demonstrated the device for Jerrold Zacharias of MIT's Radiation Laboratory, Zacharias asked if it was sturdy enough. Weber invited him to throw it on the floor, and when it did not break, Zacharias immediately ordered a thousand of them. Shortly thereafter Zacharias increased the order to ten thousand, and Weber saw the necessity of setting up a company to produce the attenuators.[27]

Thus in late 1943 PIB (Polytechnic Institute of Brooklyn) Products Company was founded. In the next several years it made attenuators of various types, each of which called for innovation in design and manufacturing. For example, when an attenuator for still higher frequencies was needed, Weber and colleagues developed an evaporative technique for depositing precisely controlled thicknesses of metal on glass. The work resulted in a score of patents, held jointly by Weber and colleagues. After the war the company, which in 1946 became Polytechnic Research & Development (PRD) Company, continued as a leading manufacturer of microwave components and instruments (see Figure 4).

Weber, who was director of PRD from its founding, also became president of the company in 1952. Running the company—its annual sales grew to $5 million—became increasingly demanding as the company grew, and Weber realized he had to make a choice between industry and academia. Since, as he says, "My heart was really in academic life,"[28] the decision was made in 1959 to sell PRD to Harris-Intertype Corporation. It continued under the name PRD Electronics as a division of that firm, and Weber worked as consultant to PRD Electronics from 1959 to 1981. The sale provided the first substantial endowment for Polytechnic.

NEBEKER: *Did PRD continue to grow after the war?*

WEBER: *Very much. We sold it in 1959 to the Harris-Intertype Corporation, and Harris-Intertype used it to build up its military-related business. They used to be a company for printing machinery, and they transformed eventually into a very large corporation, now called Harris Corporation. It's listed on the stock exchange.*

NEBEKER: *Who made the decision to sell?*

WEBER: *Well, it was really up to me as president of the company.*

NEBEKER: *I know that in 1952 you were named both president and director of PRD. But why sell? Things were going so well.*

WEBER: *Well, my heart was in teaching, and I realized that this decision comes for many people: Should they go industry or should they go academic? My heart was really in academic life. The other one was a transient phenomenon.*

NEBEKER: *You'd been doing that for quite a few years.*

WEBER: *Well, I started off with Siemens, so I had a kind of business indoctrination.*

POLYTECHNIC RESEARCH & DEVELOPMENT COMPANY, INC.

APRIL 1953

Vol. 2 No. 1

NOISE FIGURE AND SOME MEASUREMENT ASPECTS

In its common usage, noise has an undesirable connotation. It denotes sound without meaning, harmony, or rhythm. The electrical phenomenon of random fluctuation has the quality of noise if it is translated into sound waves. Hence, the term was justified when it was first coined in the early days of radio. Today we can see noise on the television screen as well as hear it.

The reduction of noise level has assumed popular significance when manufacturers of television receivers started using this as a factor in advertising. With the general acceptance of the concept of noise figure as a measure of the merit of a receiver, it is important that engineers have a knowledge of the equipment types available for measuring this important quantity.

Thermal agitation and shot noise are the two main sources of random fluctuation manifest in radio receiving equipment. Shot noise is distinguished from thermal agitation noise in that it is caused by an unidirectional stream of electrons having a random velocity distribution.

Atmospheric and man made noise are forms of disturbance which are not random in the strict sense of the term. Rather, the disturbance is sporadic and originates outside the receiving system; hence, it may be regarded as a form of external interference. Its consideration does not fall within the scope of the present discussion and will not be treated further.

It is readily appreciated that the inevitable presence of noise sets an ultimate limit to the receiver's useful sensitivity. Threshold signals, comparable in amplitude to

Figure 1. A Commercial Diode Noise Source, the PRD Type 904 VHF-UHF Noise Generator. The instrument utilizes a temperature limited coaxial type diode which permits noise figure measurements in the frequency range from 10 to 1000 mc/s. The output noise power is continuously adjustable up to a level 20 db above the noise from a reference 50 ohm resistor.

noise, are subject to its masking effect, and if the noise power exceeds that of the signal by a large enough factor, the signal will be rendered unintelligible.

In the absence of any noise contributed by the receiver, the noise power at the output of the receiver would be equal to the product of the input noise power, due to the generator or antenna radiation resistance, and the power gain of the receiver. In practice, however, the total output noise power exceeds this irreducible minimum by a factor called the noise figure, which depends on the receiver noise properties. The significance of noise figure is to be found in the fact that it fully defines an important characteristic of a receiver without reference to the magnitude of the noise present or the bandwidth of the receiver.

Noise Sources

In recent years, measurement of receiver sensitivity has been done increasingly by means of noise sources. Prior to the advent of suitable noise sources, receiver noise measurements were made with calibrated signal generators. Not only is the signal generator method of noise measurement cumbersome, but it also suffers from inaccuracies which are difficult to overcome.

The signal generator technique calls for the measurement of the noise bandwidth of the receiver. This measurement is very time consuming if reasonably accurate results are required. Furthermore, the accuracy of the low level calibration of high frequency signal generators is generally open to question. Usually, this calibration is the result of extrapolating a theoretical law of the attenuator

(Continued on page 3)

Printed in U.S.A.

Copyright 1953 Polytechnic Research & Development Co., Inc.

Figure 4. The first page of a 1953 publication of Polytechnic Research & Development Company. The photograph shows a noise source manufactured by PRD.

NEBEKER: *But you worked many years with PRD, from the early war years all the way to '59. Was it getting to be a strain running PRD, teaching, being a college administrator, and doing research simultaneously?*

WEBER: *Oh, yes. Because I had to project for the company money to*

> *operate. When they sold the company, I think we had a business*
> *volume of $5 million.*
>
> NEBEKER: *Annual sales?*
>
> WEBER: *Yes. Because eventually we had a monopoly. Hewlett Packard only*
>   *came into that with the War Department giving them all our*
>   *drawings. This is why Bill Hewlett and Dave Packard are very good*
>   *friends still.*[29]

Though dedicated to academia, Weber maintained ties to industry. In 1942 Polytechnic began offering off-campus courses at electronics companies, and in that year Weber taught linear transient analysis at the Sperry Gyroscope Company in Lake Success, Long Island.[30] Besides such work and his continuing association with PRD Electronics, he was a member of the Joint Technical Advisory Committee of the Institute of Radio Engineers and the Electronics Industry Association for most of the period from 1954 to 1962 and was a member of the New York State Advisory Council for the Advancement of Industrial Research and Development from 1959 to 1969. He was also a consultant to the military, serving on the Army Scientific Advisory Panel (1957 to 1969), the Defense Science Board (1963 to 1966), and the Army Electronics Command Advisory Group (1965 to 1970). In 1969 the US Army presented him with the Outstanding Civilian Service Award.

The research group Weber organized at Polytechnic early in the war became the Microwave Research Institute (MRI) in 1945, and Weber served as director until 1957. Weber's work in this area was well-known even before the war, and in 1941 he turned down an invitation to join MIT's Radiation Laboratory, where much of the wartime development of radar took place. He did, however, become a Rad Lab employee for a short time just after the war. This was to write up material on microwave measurements for the famous Rad Lab series of texts that were used as reference works by microwave engineers throughout the world.[31]

In the postwar years MRI became one of the leading centers of microwave research, and the research group received government recognition when Weber, in 1948, was awarded the Presidential Certificate of Merit. Between 1942 and 1956 MRI was awarded 86 research contracts with the federal government, totaling more than $5 million(see Figure 5).[32] Besides directing the institute and publishing technical papers himself, Weber organized the well-known Microwave Symposia from 1952 to 1959. These were annual international gatherings of leading microwave researchers, and the published proceedings were extremely useful as compendia of recent results. For these and other efforts, Weber was honored in 1977 with the Microwave Career Award of the Microwave Theory and Techniques Society. In 1986 MRI was renamed the Weber Research Institute.

Educator and University President

In the decade following World War II, college enrollments increased markedly and interest in electrical engineering, especially in what was bythen called electronics engineering, became intense. Indeed, in the postwar decade the number of EE graduate students in the United States increased tenfold.[33] One of the largest graduate programs in the country was Polytechnic's, which was under the direction of Weber.

Figure 5.   Weber and another member of the Microwave Research Institute observe measurements of the transmission of high-frequency waves in a waveguide.

Weber's involvement in research and in directing PRD caused no apparent diminution of his efforts as educator, and he has always taken pride in the success of his students, who include Leo Felsen, Anthony Giordano, and Nathan Marcuvitz.[34]   In the postwar decade Weber wrote two widely used textbooks. The first, *Electromagnetic Fields, Theory and Applications* (published in 1950), was a mathematically rigorous treatment that suited the needs of engineers by considering applications and by including lengthy chapters on experimental mapping methods and on graphical and numerical plotting methods. The book was so well received that it was republished in 1965.[35]   His other text was a two-volume treatment of linear transient analysis (published in 1954 and 1956).[36]   Here Weber made clear the great value for circuit analysis (particularly when high frequencies or power surges are involved) of a mastery of the underlying field concepts—that is, of Maxwell's theory.

Weber also authored a score of articles on engineering education. He has been a member of the American Society for Engineering Education (ASEE) since 1935 and was instrumental in establishing its graduate studies division. The ASEE, embracing as it does electrical, civil, mechanical, chemical, and other branches of engineering, was a welcome forum for Weber, as he has long argued for greater cooperation between the various engineering fields. After pointing to new areas of engineering such as nuclear energy, aerospace, and biomedical instru mentation and prosthetics, Weber argued, "The lines of demarcation between the engineering disciplines, so easily drawn in 1933, have become diffused and have made the plea for solidarity of the profession not only desirable but, in fact, the only sensible one."[37] Weber also has argued for a broad education for engineers: ". . . the pressure is now for even broader education beyond engineering to include the disciplines involved in the decision-making process concerning the interaction of exploitation of technology and the effects of it upon society and its environment."[38]

In 1960 Weber was awarded the Education Medal of the American Institute of Electrical Engineers "for excellence as a teacher in science and electrical engineering, for creative contributions in research and development, for broad professional and administrative leadership, and in all, for a considerate approach to human relations." Weber feels strongly the lack, in the United States, of established standards of education. Whereas the German *Maturitätszeugnis* (a certificate awarded on passing the high school final examinations) means the same regardless of the institution awarding it, an educational degree in the United States can be evaluated only with reference to the institution. Weber laments that in this country it is, for the most part, up to individual teachers to maintain educational standards.[39]

The active role Weber assumed in institutional affairs at Polytechnic Institute, where he demonstrated an ability to organize collective efforts and to motivate individuals, led to his appointment in 1957 as the first vice president for research; later that year he became president (see Figure 6). In his inaugural address he expressed his educational philosophy: "No longer can the engineer just be equipped with skills, as it appeared desirable even twenty-five years ago. Today we must educate *scientific engineers*, well-founded in the fundamental laws of science and able to keep pace with new scientific developments."[40]

Weber served as president of Polytechnic from 1957 until June 1969, when he was elected president emeritus and professor emeritus. During those years, Polytechnic underwent a remarkable growth in university facilities, in academic programs, and in number of students. Soon after assuming the presidency, Weber directed the move to a new campus facing Jay Street in Brooklyn. A further expansion followed in the early 1960s when a campus exclusively for research and graduate studies was opened in Farmingdale, Long Island. Besides affording Polytechnic much needed space, the new facility made possible even closer ties between the university and industry, since many high technology companies were located nearby. Through an Industrial Research Associates Program, Polytechnic made consulting services, conferences, and seminars available to companies in the area; a Continuing Professional Studies Program was established to offer instruction—in intensive two-week sessions and in short courses—to company executives and engineers.

Weber's principal goal as president was to make Polytechnic a leading university for both research and graduate education in engineering and science. The success of this effort is indicated by the growth of the graduate programs. While the number of bachelor's degrees awarded annually remained fairly constant, the number of master's degrees increased from 150 in 1958 to 470 in 1970 and the number of doctoral degrees increased from 36 in 1958 to 108 in 1970. In the same period, the number of Ph.D. programs increased from seven to sixteen.[41] The success is indicated also by the high standing the university had achieved by the time Weber

Figure 6. Weber's investiture as president of Polytechnic Institute of Brooklyn by Preston R. Basset, chairman of the Board of Trustees.

stepped down as president: A study conducted by the American Council on Education rated the graduate EE program at Polytechnic as sixth in the nation in quality of faculty and eighth in effectiveness of program.[42]

NEBEKER: *I know you had a lot of Japanese students.*

WEBER: *Yes. We had students from all over the world, in fact.*

NEBEKER: *How do you explain that?*

WEBER: *Because of our leadership during the war in radar. Also our symposia that started in '52. They helped a tremendous amount. . . .*

NEBEKER: *How did finances go in those years for Poly?*

WEBER: *Well, we established that Alumni Fund where alumni contributed rather substantially. . . .*

NEBEKER: *And you were able to expand in those years?*

WEBER: *Yes.*

NEBEKER: *Of course in those years you were also very active in professional societies, president of IEEE, for example. It must be very natural for someone to say, "Okay, now I just can't continue doing all these different things. I'll have to concentrate my energies."*

WEBER: *I just felt there are so many opportunities to do something that I wanted to use some of them. I probably shouldn't say this, but a member of the faculty called my period as president the* Blütezeit, *meaning the bloom period. So although I was president of the faculty, I knew them well enough to know both their strengths and their weaknesses. So this gave me greater power.*

NEBEKER: *To strengthen the faculty?*

WEBER: *Yes, we got very excellent faculty members at that time.*[43]

Contributor to the Science and the Profession of Electrical Engineering

Since his days at Siemens when he used conformal mapping and Maxwell's equations to solve problems of practical engineering, Weber has advocated the greater use of mathematics and physics in engineering. In 1937, perceiving a great interest in physics among members of the American Institute of Electrical Engineers (AIEE), Weber organized the Basic Science Group of the New York AIEE Section, which sponsored a series of lectures aimed at bringing more physics into engineering.[44] Weber himself maintained ties with the physics community—he joined the American Physical Society (APS) in 1931 and was named an APS Fellow in 1946. He was also for many years a member of the American Mathematical Society, and he has been active in the New York Academy of Sciences and the American Association for the Advancement of Science.

Another of Weber's abiding concerns was for the units and standards of electrical engineering. When he was working in Berlin, Weber was an active member of the *Ausschuss für Einheiten und Formelgrössen* (Commission on Units and Standards). He had already published two articles on electrical units when he was asked to write the section entitled "Physical units and standards" of the widely used *Handbook of Engineering Fundamentals* (1936).[45] His contributions in this area were honored in 1966 with the award of the Howard Coonley Medal of the American Standards Association.

Related to this concern was Weber's work on techniques of measurement. In his investigations of microwaves, he built attenuators, wave meters, and frequency meters. Indeed, shortly after the outbreak of World War II he delivered an attenuation standard for microwaves to the National Bureau of Standards. In both the AIEE and the Institute of Radio Engineers (IRE) he served as chairman of instrumentation and measurement committees.

Weber has contributed to his profession in many other ways. Since joining the *Österreichischer Verband für Elektrotechnik* (Austrian Association of Electrical Engineers) in 1923, Weber has been active in professional organizations. He joined the AIEE in 1931, was named a fellow in 1934, and served on many committees. He played an even larger role in IRE affairs, including ten years as a director (1952–1962) and terms as president (1959) and vice president (1962).[46] Like their successor (IEEE), the AIEE and the IRE were largely volunteer organizations, which succeeded only because members willingly gave of their time and energy. Weber, to an unusual degree, felt a social obligation to his professional organizations:

> Any social institution *is* what its members make it! If its members do not care, if they let "the others" worry, let "the others" be the fools to work—then that institution, however great it might appear at the moment, is *doomed*; it will crumble and disappear. If, on the other hand, there are enough members who *actively* support the institution, who freely devote time and effort to its tasks and obligations, who share in the belief of a common goal and of common interests in higher achievements, then that institution will grow strong and prosper.[47]

Weber also felt strongly that engineers are part of an international community, and he personally maintained a great many international ties.[48] (See Figure 7.) He was, for many years, a member of the US National Committee of the International Union of Radio Science (URSI), and is an honorary life member of URSI. In the late 1930s and early 1940s, Weber assisted Jewish refugees in finding positions in the United States. He regretted that Polytechnic was not then able to create new positions: "If I had had at that time real money, I could have built up a faculty unequaled anywhere."[49] In 1961 he was one of five past presidents of IRE sent to Europe to visit existing sections and create interest in new ones.[50]

On the first day of 1963 the Institute of Electrical and Electronics Engineers came into existence by the merger of the AIEE and the IRE.[51] These were two large organizations—AIEE had almost 70,000 members, IRE over 100,000, and together they published 39 journals—having different procedures and traditions. Their amalgamation was viewed nervously by many members. The joint merger committee "agreed that the first president would set the tone to be followed" and selected Ernst Weber for the position. John D. Ryder, one of the IRE representatives on the committee, said, "I don't remember any other name ever being discussed."[52]

Figure 7. Weber lecturing during a 1954 visit to France. Weber, who could speak French, gave talks in France on several occasions.

In his year as president, Weber played a large part in the countless compromises necessary to make one organization of two. He had long been active, and was held in high regard, in both AIEE and IRE, and he had the skills of a mediator required for that difficult job. Clarence Linder, one of the AIEE representatives on the merger committee, recalled that Weber's "nature was to draw the parties together and try to have them work out their differences...."[53] In 1963 Weber visited more than 40 of the 120 IEEE Sections, always conveying the feeling that the merger would be successful. He also made a two-month trip (at his own expense) around the world to stimulate interest in forming IEEE Sections abroad; in this he was helped by his many former students, quite a few of whom had attained positions of prominence in their native countries. Indeed, in Japan, Polytechnic had an alumni group.[54] Weber remained on the IEEE Board of Directors until 1965, and in 1971 he was honored with the IEEE Founders Award "For leadership of great value to the profession."

Weber has served his profession in other ways. He was a member of the New York Electrical Society from 1941 to 1954 and president in 1946 and 1947. From 1964 to 1972 he served on the board of directors of the Engineers' Council for Professional Development, and from 1968 to 1970 he was president of the Council. In 1978 he was honored by the Engineers' Council for Professional Development with the L.E. Grinter Award.

In 1964 Weber was named a founding member of the National Academy of Engineering, and he was also elected to the National Academy of Science. These two academies serve both the government and society in general by providing scientific and technical advice through the National Research Council (NRC). On Weber's retirement from Polytechnic in 1969, he was named chairman of the NRC's Division of Engineering. He agreed to give seventy percent of his time to this job and leased an apartment in Washington, which became his primary residence for the next nine years. He oversaw the NRC investigations of such issues as standards for fire safety, the possibility of earthquake prediction, and motor vehicle pollution. His constant concern was to see that all views were considered, and he was especially cognizant of the danger of industrial bias within investigative committees. After five years as chairman of the Division of Engineering, he served four years, part of the time as acting director, on the NRC's Commission on Sociotechnical Systems.

In this work Weber was following his own precept that "the engineer must assume a greater responsibility in the socioeconomic and political spheres of society...."[55] In an article entitled "The engineer's responsibility to society," Weber quoted J. Douglas Brown, dean of the faculty at Princeton University: "... the engineer must be cognizant of the needs and aspirations of mankind while interpreting and implementing the knowledge science has discovered. The professional engineer must... be a bridge between the two worlds of science and humane values."[56] Weber was doing just this when, during his tenure as president of Polytechnic, he worked to establish at the university a Center for Urban Environmental Studies, which sought to improve urban transportation, to reduce air pollution, to optimize the structure of local government, and to upgrade urban communications systems.[57]

Bridger of Cultures

Through most of his career, Weber had the constant support of his wife Sonya. His first marriage, to Irma Linter, who came from the same neighborhood in Vienna, ended after eight years in 1933. Three years later he married Charlotte Sonya née Escherich, who had two daughters by a previous marriage. Sonya, also a native of Vienna, was the daughter of Theodor Escherich, a pediatrician and bacteriologist famous for the discovery of the intestinal bacteria named after him, *Escherichia coli*. Sonya's earlier marriage was to Hugo Eisenmenger, an Austrian engineer who had settled in the United States.

Sonya Weber became well known as a physiotherapist—she had earned a doctorate in physiotherapy in 1934—and was for many years a director of the children's clinic of the Presbyterian Hospital in New York City. She was in demand as a lecturer on physical fitness, and Ernst was in demand as speaker at IRE and AIEE meetings, so on many occasions the couple traveled together to give lectures in the same town. On the occasion of the establishment of the Sonya and Ernst Weber Scholarship Fund in 1975, James Flack, son-in-law of Sonya, said, "This scholarship fund in their names together is most fitting. They live and give of themselves as one. They have the same objectives, the same high standards. Each is the other's greatest supporter."[58] Sonya, who died in 1984, is lovingly remembered in a book that Weber wrote entitled, *A Brief History of the Family of Sonya Escherich-Eisenmenger-Weber*.[59]

Ernst and Sonya's acquaintance came about through a chance meeting, and it was also a chance meeting that led to Weber's invitation to come to the United States. Reflecting on these and other events in his own life, Weber has commented that few people sufficiently appreciate how much their lives are shaped by chance. The truth of that notwithstanding, it is certainly the case that Weber, like a Horatio Alger hero, benefited from many chance meetings partly because he made a very favorable impression on people. Colleagues describe him as a perfect gentleman, forward-looking and wise, holding to the highest standards of personal integrity and exercising a quiet leadership.

Weber's productive career as practicing engineer, researcher, educator, and leader in his profession has been abundantly honored. To the many distinctions already mentioned may be added his election to the American Academy of Arts and Sciences and his receipt of honorary doctorates from six institutions.[60] For his pioneering research in electromagnetic fields, linear and nonlinear circuits, and microwave measurements, Weber was awarded the National Medal of Science by President Reagan in 1987.

Throughout his life Weber has worked to build bridges between different cultures: between European electrical engineering and American electrical engineering, between industry and academia, between engineering and physics, and between the AIEE and the IRE. He has also provided a bridge from past to present with his longevity and continued activity and with his appreciation of the achievements of the past, manifested in his historical writing.[61] What is more, he has provided a bridge to the future through his students and the hundreds of others deeply influenced by him and through the professional and educational institutions that he helped sustain and give new life. According to Anthony Giordano, a colleague and friend of Weber's for over sixty years, Weber is exceptional in always looking to the future, searching out the potential of individuals and institutions and seeking ways to realize that potential, and in motivating people by providing them with an optimistic outlook on life that they did not have before.

---

[1] *The New Encyclopaedia Britannica* (1991), vol. 7, p. 969. The quotation is from Michael Shaw's translation of *Winnetou* (New York: Seabury Press, 1977), p. 4.

[2] Suzanne Tyndel, "Karl May," in Walther Killy, ed., *Literatur Lexikon* (Gütersloh: Bertelsmann Lexikon, 1990), vol. 8, pp. 26–28, and Roland Smith, "Günter de Bruyn and *Neue Herrlichkeit*," in Arthur Williams, et al. (eds.), *Literature on the Threshold: The German Novel in the 1980s* (New York: Berg Publishers, 1990), pp. 77–90.

[3] The information about Ernst Weber contained in this article comes mainly from the following sources: (1) an extensive oral history interview of Weber conducted by the author 11–12 April 1991 (from which an edited transcript has been prepared); (2) Weber's published writings; (3) copies of personal papers provided by Weber to the author; and (4) the tapes and transcript of an interview of Weber conducted by Trudy Bell and Don Christiansen on 9 March 1988. (Items 1, 3, and 4, as well as a full list of Weber's publications, are available at the IEEE Center for the History of Electrical Engineering.) For assistance with this and other chapters of the book, the author would like to thank William Aspray and Andrew Goldstein. An earlier version of this article appeared in *Proceedings of the IEEE*, vol. 81 (1993).

[4] Interview 1991, p. 34.

[5] Interview 1991, p. 11.

[6] Interview 1991, p. 12.

[7] Ernst Weber, "Historical notes on microwaves," *Proceedings of the Symposium on Modern Advances in Microwave Techniques* (New York: Polytechnic Press, 1954), vol. IV, pp. 1–23.

[8] Ernst Weber, "Oliver Heaviside—biography," introduction to Oliver Heaviside's *Electromagnetic Theory* (New York: Dover Publications, 1950), pp. xv–xvii.

[9] Interview 1991, pp. 16–17.

[10] Sigfrid von Weiher and Herbert Goetzeler, *The Siemens Company—Its Historical Role in the Progress of Electrical Engineering 1847–1980*, second English edition (Berlin: Siemens, 1984), p. 42; Sigfrid von Weiher, "Reinhold Rüdenberg," *Dictionary of Scientific Biography*, vol. XI (1975), pp. 588–589. Rüdenberg studied electrical and mechanical engineering at the Technische Hochschule in Hannover, where he earned a doctorate in 1906. After a

period as assistant to Ludwig Prandtl in Göttingen, he went to work in 1908 for Siemens-Schuckert in Berlin. His more than 300 patents speeded the development of engineering practice, and his more than 100 publications contributed to the establishment of the science of electrical engineering. His textbook on electrical switching processes (first edition, 1923; fourth edition, 1950) is a classic.

[11] P. Rothe, ed., *Funktionentheorie und ihre Anwendung in der Technik* (Berlin: J. Springer, 1932). The English translation was published by MIT Press in 1935.

[12] Interview 1991, p. 29.

[13] Ernst Weber, *A Brief History of the Family of Sonya Escherich-Eisenmenger-Weber* (Tryon, NC: M.A. Designs, 1990), p. 69, and Interview 1991, p. 22.

[14] Interview 1991, p. 38.

[15] See Ernst Weber's "The future role of graduate study in engineering" in *Journal of Engineering Education*, vol. 45 (1954), pp. 236–239; and Chapter 6, "The new world of electronics engineering," in A. Michal McMahon's *The Making of a Profession: A Century of Electrical Engineering in America* (New York: IEEE Press, 1984).

[16] John D. Ryder, "The way it was," *IEEE Spectrum*, vol. 21, no. 11, 1984, pp. 39–43.

[17] Ernst Weber, "Types of graduate subsidy and their relation to educational values," *Journal of Engineering Education*, vol. 44, 1953, pp. 188–191.

[18] Interview 1991, p. 31.

[19] Ernst Weber, "The challenges in the development of graduate programs," *IRE Transactions on Education*, vol. E-2, 1959, pp. 39–43.

[20] Ernst Weber, "The future role of graduate study in engineering," *Journal of Engineering Education*, vol. 45, 1954, p. 238.

[21] Ernst Weber, "Types of graduate subsidy and their relation to educational values," *Journal of Engineering Education*, vol. 44, 1953, pp. 188–191.

[22] Interview 1991, p. 68.

[23] Interview 1991, pp. 46–47.

[24] "Electrical engineering at Polytechnic" in *State-of-the-Art and the Future of Electrical Engineering* (program for "A Symposium Celebrating the Centennial of Electrical Engineering at Polytechnic University and Honoring Dr. Ernst Weber on his 85th Birthday, September 22–24, 1986").

[25] Ernst Weber, "The challenges in the development of graduate programs," *IRE Transactions on Education*, vol. E-2, 1959, pp. 39–43.

[26] "Ernst Weber," autobiographical article in *McGraw-Hill Modern Scientists and Engineers* (New York: McGraw-Hill, 1980), vol. 3, pp. 283–284.

[27] Most of the information in this and the preceding paragraph comes from an unsigned article entitled "A story of one of Poly's wartime research programs—attenuators" in the July 1946 issue of *Poly Men* and from a lengthy press release from Polytechnic headed "for release Saturday P.M.,

June 8th and Sunday A.M., June 9th" (no year is given, though it was probably 1946).

[28] Interview 1991, p. 58.

[29] Interview 1991, pp. 66–67.

[30] "Polytechnic's educational programs in cooperation with industry" in *State-of-the-Art and the Future of Electrical Engineering* (cited above).

[31] Chapter 12 ("Resistive attenuators"), coauthored with R. N. Griesheimer, and Chapter 13 ("The measurement of attenuation"), of *Technique of Microwave Measurements*, vol. 11, Radiation Laboratory Series (New York: McGraw-Hill, 1947).

[32] *Report from Polytechnic Institute of Brooklyn*, vol. 1, no. 3, February 1956.

[33] Ernst Weber, "The challenges in the development of graduate programs," *IRE Transactions on Education*, vol. E-2, 1959, pp. 39–43.

[34] Leopold B. Felsen, winner of the IEEE Heinrich Hertz Medal and member of the National Academy of Engineering, was named a Fellow of the IRE "For contributions to electromagnetic theory and measurement." Nathan Marcuvitz, who also is a winner of the Heinrich Hertz Medal and member of National Academy of Engineering, was named an IRE Fellow "For fundamental contribution to the solution to microwave field problems." Anthony B. Giordano, honored as an IRE Fellow "For his contributions to microwave measurements and modern communication curricula," has served as a Director of IEEE and as President of the American Society for Engineering Education.

[35] *Electromagnetic Fields—Theory and Applications*, vol. 1: *Mapping of Fields* (New York: John Wiley, 1950); republished as *Electromagnetic Theory: Static Fields and Their Mapping* (New York: Dover, 1965).

[36] *Linear Transient Analysis*, 2 vols. (New York: John Wiley, 1954 and 1956); volume 1 is subtitled *Lumped-Parameter Two-Terminal Networks*, and volume 2 is subtitled *Two-Terminal-Pair Networks, Transmission Lines*.

[37] Ernst Weber, "The engineer and society," *Electrical Engineering*, vol. 82, 1963, pp. 438–441.

[38] Ernst Weber, "Science and societal engineering," *Proceedings of a Symposium on Submillimeter Waves* (Brooklyn, NY: Polytechnic Press, 1971), pp. xiii–xix. See also Weber's "Technological challenges to educating engineers," *IEEE Spectrum*, vol. 2, 1964, pp. 119–120.

[39] Interview 1991, p. 83.

[40] "Inauguration and dedication ceremonies, April 19, 1958," Polytechnic Institute of Brooklyn.

[41] Most of the information in this and the preceding paragraph comes from a 12-page manuscript, "Dr. Weber's contributions as President of Polytechnic" (1992) by Anthony B. Giordano, a copy of which is available at the Center for the History of Electrical Engineering.

[42] Allan M. Cartter, *An Assessment of Quality in Graduate Education*

(Washington, DC: American Council on Education, 1966), pp. 74–75. Polytechnic was also ranked nationally in chemistry (faculty "Strong," program "Acceptable plus"), physics (faculty "Adequate plus," program "Acceptable plus"), chemical engineering (faculty "Good," program "Acceptable plus"), and mechanical engineering (faculty "Good," program "Acceptable plus"). (Faculty were classified as Distinguished, Strong, Good, Adequate Plus, or "not grouped"; programs were classified as Extremely Attractive, Attractive, Acceptable Plus, or "not grouped.")

[43] Interview 1991, pp. 87–88.

[44] For example, in the winter of 1941/42 the Basic Science Group presented a series of lectures by six different engineers, including Weber, on nonlinear circuit theory, and the following year Weber gave all six lectures in a series on ultrashort electromagnetic waves.

[45] Ovid W. Eshback, ed., *Handbook of Engineering Fundamentals* (New York: John Wiley, 1936). Weber's earlier articles were "A proposal to abolish the absolute electrical unit systems," *Transactions of the AIEE*, vol. 51, 1932, pp. 728–742, and "Ein Vorschlag zur Lösung des Problems der Electrischen Einheiten-systeme," *Elektrotechnik und Maschinenbau*, vol. 50, 1933, pp. 45–51. Weber was also the author of the section on radiation and light in *Handbook of Engineering Fundamentals*, and Weber prepared revised versions of his two sections for the second edition of this handbook, which appeared in 1952.

[46] He was a Director from 1952 to 1962, President for 1959, and Vice President for 1962. A listing of his service on IRE committees is impressive: AdCom of Board of Editors 1951; Annual Review 1950–51; Appointment 1959–61; Awards 1952–54, Chairman 1953–54; Awards, Coordination 1958; Board of Editors 1949–53; Circuits 1949–53; Editorial Administrative 1949–51; Education 1944–51; Executive 1957–61, Chairman 1959–60; Finance 1952, 1959–60; Instruments and Measurements 1948–52, Chairman 1949–51; Measurements and Instrumentation, Chairman 1945–55; Membership 1941–48; Chairman, Technical Program Committee, National Convention 1947, 1951; Nominations 1959–61; Policy Advisory 1952, 1956, Chairman 1960; Policy Study, Chairman 1960; Professional Groups 1951–60, Vice Chairman 1954–56, Eastern Division Vice Chairman 1957–59, Eastern Division Chairman 1960–62; Standards 1949–60, Vice Chairman 1951–54, Chairman 1954–56; Ex Officio 1957–59.

[47] Ernst Weber, "Why awards?" *Proceedings of the IRE*, vol. 39, 1951, p. 595.

[48] In 1937 Weber published an article in *Scientific Monthly* (vol. 44, pp. 171–173) on "The international mission of science" in which he argued that science promotes internationalism, not by propagandistic methods, but by engaging people in an international enterprise that requires keeping abreast of research worldwide and conduces to admiration of achievements made in other countries.

[49] Interview 1991, p. 55.

[50] Trudy Bell, "Piloting the IEEE through a critical first year," *IEEE Spectrum*, vol. 25, 1988, no. 10, pp. 42–44.

[51] The merger, which began more than a decade earlier, is well described in Chapter 12, "AIEE + IRE = IEEE" (pp. 209–231) of John D. Ryder and Donald G. Fink's *Engineers & Electrons: A Century of Electrical Progress* (New York: IEEE Press, 1984); in the section entitled "The path to merger: The founding of the Institute of Electrical and Electronics Engineers" (pp. 239–243) of A. Michal McMahon's *The Making of a Profession*; and in Trudy Bell's "Piloting the IEEE. . . ." The formation of IEEE is well documented in the archives of the Center for the History of Electrical Engineering; more than 1000 documents—meeting minutes, committee papers, legal documents, correspondence—deal with the merger itself.

[52] Quoted in Trudy Bell's "Piloting the IEEE. . . ."

[53] Ibid., p. 42.

[54] In 1963 Weber was named an honorary member of the Institute of Electrical Engineers of Japan and of the Institute of Radio Engineers of Japan.

[55] Ernst Weber, "The engineer and society," *Electrical Engineering*, vol. 82, 1963, pp. 438–441.

[56] Quoted in Ernst Weber, "The engineer's responsibility to society," *Michigan Quarterly Review*, vol. 4, 1965, pp. 206–211.

[57] Giordano, "Dr. Weber's contributions as President of Polytechnic."

[58] In Weber, *A Brief History of the Family of Sonya Escherich-Eisenmenger-Weber* (cited above), p. iv.

[59] Privately printed by M.A. Designs, Tryon, NC, in 1990.

[60] Sc.D., 1958 Pratt Institute; D.Eng., 1959 Newark College of Engineering; Sc.D., 1959, Long Island University; L.L.D., 1963, Brooklyn Law School; D.Eng., 1964, University of Michigan; and D.Eng., 1970, Polytechnic Institute of Brooklyn.

[61] These include "Historical notes on microwaves," *Proceedings of the Symposium on Modern Advances in Microwave Techniques*, vol. IV, 1954, pp. 1–23; "The engineer and society" *Electrical Engineering*, vol 82, 1963, pp. 438–441, which is a review of the history of the engineering profession; a biography of Oliver Heaviside, published in a 1950 reprint of Heaviside's *Electromagnetic Theory*; and the book, mentioned above, on his wife's family. He is presently at work on a book tentatively entitled *A Global Perspective on the Early Evolution of Electrical Engineering*.

CHAPTER 2

Working to Establish a New Discipline
Herman P. Schwan
and the Roots of Biomedical Engineering

HERMAN P. SCHWAN

Figure 1. Herman P. Schwan, winner of the IEEE Edison Medal, member of the National Academy of Engineering, recipient of an honorary doctorate from the University of Pennsylvania, and current resident of Radnor, Pennsylvania.

Biomedical engineering—the use of the principles and techniques of engineering to solve problems in biology and medicine—has roots deep in history. Luigi Galvani's investigations in the 1780s of "animal electricity" began a line of research, called electrophysiology, that before the end of the nineteenth century had established the electrical nature of the nerve impulse and revealed much about electrolytic conduction in animal tissues. Here concepts of electrical engineering—such as resistivity, capacitance, and polarization—were obviously applicable. Indeed, the mathematical model William Thomson (later Lord Kelvin) proposed in 1855 for understanding the Atlantic telegraph cable was imported intact (see Figure 2).

Where electrical engineering had a much greater effect on biology and medicine was in instrumentation. By 1910 thermocouples, capillary electrometers, string galvanometers, and many other electrical devices were commonly used in biomedical research.[1] Electronic devices also found early application: x-ray imaging dates from 1895; both cathode-ray tubes

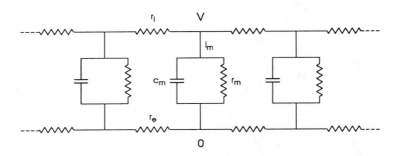

$$\frac{\partial^2 V}{\partial x^2} = \frac{r_e + r_i}{r_m} V + (r_e + r_i) \, c_m \, \frac{\partial V}{\partial t}$$

Figure 2. A schematic representation of a cable and an equation describing its electrical behavior, based on William Thomson's 1855 analysis of an underwater telegraph cable. By 1905 this analysis was being used to describe conduction along a nerve fiber.[2] (V is electrical potential difference across the membrane (or insulating sheath), x is distance, R_e is external resistance per unit length, R_i is internal resistance per unit length, R_m is membrane resistance per unit length, C_m is membrane capacitance per unit length, and t is time.)

and vacuum-tube amplifiers were being used to study nerve impulses in the 1920s; and the electron microscope, invented in 1931, was being used in biology by 1934.[3]

Not only research, but also medical practice was changed. Many research instruments, such as x-ray machines, electrocardiographs, and electroencephalographs, came to be regularly used in diagnosis. Throughout the nineteenth century there were attempts to use electrical technology in treatment but only a few (such as cardiorespiratory resuscitation by electrical stimulation) were of much effectiveness.[4] In the early decades of this century, a few techniques of obvious utility were introduced: x-ray therapy, electrosurgery, and diathermy (the generation of heat in animal tissues by electromagnetic or ultrasonic radiation). In the late 1920s, the vacuum tubes developed for radio led to short-wave diathermy, electrosurgery, and medical applications of telemetry.[5]

In the 1920s and 1930s, as more and more investigators used the concepts and techniques of physics and engineering in biological and medical research, a few institutions were established to promote this approach. In the United States in the 1920s, the Johnson Foundation for Medical Physics at the University of Pennsylvania and the Biophysics

Department of the Cleveland Clinic were founded. Siemens, a major supplier of x-ray and diathermy equipment, maintained a biophysical laboratory at Erlangen.[6]

Also in the 1920s the Oswalts, a Frankfurt family that had become wealthy through banking, provided the money for an Institute for the Physical Foundations of Medicine (*Institut für physikalische Grundlagen der Medizin*). The founding director was Friedrich Dessauer, who is best known for his theory of x-ray damage. In 1934 Boris Rajewsky became director of the institute; he studied the biological effects of both ionizing and nonionizing radiation.[7] In 1938 Rajewsky gained the sponsorship of the Kaiser Wilhelm Society, and the Oswalt Institute for the Physical Foundations of Medicine was attached to the larger, newly formed Kaiser Wilhelm Institute for Biophysics (*Kaiser Wilhelm Institut für Biophysik*, later the *Max Planck Institut für Biophysik*). Both biomedical engineering and biophysics—indeed, it is often difficult to distinguish between these disciplines—advanced at the Institute, where the main areas of research were the biological effects of electromagnetic radiation, both ionizing and nonionizing, and dosimetry of x-rays. With the new funding from the Kaiser Wilhelm Society, the Institute increased its staff to about twenty employees. Among those hired as a technician in 1937 was a young engineer who reached that position by a tortuous path.

Growing Up in Weimar and Nazi Germany

Herman Schwan was born 7 August 1915, in the summer of the Second Battle of Ypres, the Gallipoli Campaign, and US protests over the sinking of the *Lusitania*.[8] His father, Wilhelm Schwan, having recovered from a serious injury suffered early in the war, was teaching mathematics and physics in the high school of Aachen, a city near the border to Belgium and Holland. His mother, Meta née Pattberg, was from a well-to-do family: Her father was director of the railway system in Westphalia. In 1918 the Schwans moved to Bad Kreuznach, a town near Mainz, where the Main River joins the Rhine. Wilhelm Schwan taught at the high school, wrote a widely used geometry text, and edited a volume of mathematics lectures.[9] He became affiliated with the mathematics faculty of the University of Frankfurt and received his doctorate there.

Wilhelm Schwan, a supporter of the Social Democrats, was outspoken in his opposition to the militant nationalism then finding expression, and he strenuously objected to the view that force should be used to correct the "injustice" of the Versailles Treaty that ended the World War. As time passed, his political views became increasingly unpopular. He and three other teachers with liberal views were socially ostracized, and the atmosphere became intensely unpleasant for them. One of the teachers committed suicide. Another emigrated to France. Wilhelm Schwan was helped out of this situation by friends who, in 1926, arranged for his transfer to the high school in Düsseldorf. Three years later, however, Wilhelm was again transferred, this time to Meseritz in eastern Germany.

On 30 January 1933 Hitler was appointed Chancellor, and on 7 April the Nazi government enacted the "Law for Restoration of the Professional Civil Service," which led to the dismissal from government employment of Jews and of people not holding "proper" political beliefs. Among them was Wilhelm Schwan, and he was unable to find other employment. He became severely depressed and had periods of imagining that the Gestapo had placed microphones in the walls of the family's apartment. He moved to Berlin.[10]

Several years before this, Wilhelm and Meta Schwan had separated, though they were reunited briefly in Göttingen, where Meta, with 15-year-old Herman, had taken up residence. She had selected Göttingen primarily because of its excellent high school and university. In the years Herman attended the high school (1930 to 1934), the children of Hermann Weyl, Max Born, and Richard Courant were also students. According to Herman Schwan, ". . . the Göttingen years were very, very important to me. I had an excellent school with excellent teachers. There was a very intellectual environment. . . ."[11] The influence of his father was no doubt partly responsible for Schwan's strong interest in mathematics and physics. Equally strong was his interest in history; besides taking many history courses, Schwan avidly read ancient and Russian history as well as the history of the World War.

In 1934 Schwan received his *Abitur*, graduating summa cum laude. At that time entry into a university required not only the *Abitur*, but also a certificate of political maturity (*Reifezeugnis*). The latter was conferred by Nazi officials, and Schwan, whose political views were known to teachers from discussions in history classes, did not receive the certificate. Much against his inclination, Schwan then entered the Reich Labor Service (*Reichsarbeitsdienst*) as a way of proving "political maturity." The Labor Service provided quasimilitary training to large numbers of Germans without contravening the restriction on the size of the military mandated by the Versailles Treaty, and it soon became obligatory for all young men.

Schwan describes the "slave labor camp" he found himself in: "Getting up at four in the morning for exercising and singing patriotic songs. . . . Then we had to march to where we were building an airfield which became a Stuka base."[12] There it was backbreaking work until two in the afternoon, followed by the march back to the main camp and two hours of military exercises. After the evening meal, political indoctrination lasted until lights-out at ten. He soon developed major health problems: "I just couldn't take that physical punishment."[13] So just six weeks after entering the Labor Service, Schwan was granted a discharge by the camp doctor and received the political maturity certificate.

This was not, however, the last hindrance to university study, as it was necessary for Schwan to receive a tuition waiver. Advised that because of his political views he had no chance of receiving such an award at Göttingen, Schwan applied to the University of Frankfurt (where his father was well known to the math faculty) and was granted the tuition waiver. The award, however, required his membership in the "Comradeship House," which was Nazi dominated. Because of persecution at the House (see quotation below), he asked to leave it, and this meant loss of his tuition waiver. So Schwan returned to Göttingen, where his mother was still living, for his second year of study. She and he saved and borrowed all they could to pay the tuition at the university there. "Somehow we did it . . . but after one year there it was all out. There was no chance to continue [at Göttingen]."[14]

. . . I fell in disgrace, unfortunately, fairly rapidly with the people in the so-called Comradeship House. It must have been that I was not too careful about expressing myself politically, even though I tried to be careful. It became known that I was invited to Sunday dinners several times by some of my professors who were Jewish. I interacted with them, of course, strongly.

. . . It just happened that I was the best in my class. When they asked a question of the class, I raised my hand first. I did some things which others didn't. For example, sometimes before class started when the blackboards were all dirty, I wiped them clean before the professor came. Well, they called me "Der Judenjüngling." *How can I translate that? It means "the Jew boy." I became known as a Jew boy. One night they beat me up. It was a gruesome experience which led again to my heart deficiencies. I was surrounded by a blanket or sack, and they beat on me with sticks. It was an awful experience. I was deep in sleep when it started. It was one of the most miserable things in my life. In the morning I had all sorts of heart problems. I was permitted to leave the Comradeship House. That meant the next year I couldn't get free tuition.*[15]

Schwan soon devised a way to continue his education. First, he would earn money at a summer job as an engineer, then he would study at a university in the eastern part of the country where, he had heard, scholarships were easier to obtain because the government wanted to encourage ethnic Germans to live in areas where there were large Polish-speaking populations.

The first step went as planned. Schwan had done very well in course work, both in Frankfurt and in Göttingen. In addition to mathematics and physics, he studied engineering, a subject his hobby of building radios had made attractive to him. So he was able to get a summer job at the renowned firm of Siemens and Halske in Berlin, where he worked on developing electric hygrometers.

The second step failed. Schwan did move to Breslau (today part of Poland and called Wroclaw) in Lower Silesia and began attending the university. It took some time to learn if he would receive a scholarship, and after one semester he learned he would not. "I remember . . . I was in my room sick. I had a vicious flu. I hadn't paid my rent for two months, and the owner of the apartment was threatening to evict me. . . . I got a note from the university that I was thrown out for not paying tuition."[16]

His hopes for a university degree seemingly come to naught, Schwan took a job as engineer at Telefunken, which was the largest German manufacturer of radio receivers and the principal German center for research in high-frequency techniques. In the main Telefunken facilities in Berlin he tested radio receivers; he recalls evaluating an RCA receiver, comparing it to Telefunken receivers.

One day, about half a year after starting at Telefunken, Schwan had a chance encounter in the company cafeteria with one of his physics professors at Frankfurt. Schwan explained what he was doing there and why he had stopped his studies. The professor said it was a shame that so promising a student was not continuing his education and, without Schwan's knowledge, contacted H. Daenzer, another professor who knew Schwan well. Daenzer made inquiries at Frankfurt, and some weeks later Schwan received an offer of a job at the Institute for the Physical Foundations of Medicine. He would be hired as a technician at a low salary, but he would be allowed to take courses at the university and tuition would be paid.

Though delighted to be able to continue his studies, Schwan was also reluctant to leave Telefunken. He decided to maintain ties with the company so that after completing his Ph.D. he might return to work there. He therefore arranged with the director of the Institute in Frankfurt to be given a quarter of the year free in order to work at Telefunken. Indeed, the following summer (1938) Schwan did return to Telefunken and was assigned to the high-frequency laboratory where he worked on high-frequency power transmission. (In the same laboratory was a magnetron development group, which was disbanded by government order in 1939 as being inessential to the war effort. In 1943, as the importance of radar and the relative backwardness of the Germans in this area became clear to Nazi authorities, the group was reestablished.)

It was in October 1937 that Schwan moved to Frankfurt and joined the Institute. There he worked with a small group, led by the director of the institute, Boris Rajewsky, studying the electrical properties of biological tissues and the technique of diathermy. Schwan saw immediately that the Institute's electrical and electronic instruments—oscillators, Wheatstone bridges, and other devices—were at quite a primitive level, and knew that one could achieve greatly increased accuracy. Rajewsky quickly recognized Schwan's abilities and supported his research initiatives. At the same time Schwan was taking courses, including several on biophysics taught by Rajewsky, several physics laboratory courses, and a demanding course in

analytic techniques taught by the famous mathematician Carl Ludwig
Siegel (who shortly thereafter emigrated to the United States).

World War II and Project Paperclip

The German invasion of Poland in September 1939 brought on world war,
but for about two years Schwan's situation changed little. Living conditions in Frankfurt were not greatly affected—there was rationing of basic
foodstuffs, but this had begun before the war in accord with Hermann
Göring's slogan "guns before butter"—and research at the Institute and
teaching at the University continued much as before. Schwan gave most
of his time to what he had selected as his dissertation research: deciding
between two alternative theories to explain the high-frequency properties
of biological tissues. He also collaborated with H. Schaefer, also at the
Institute, to disprove a theory proposed by the Russian researcher N. N.
Malov to use electric fields to obtain large, localized elevations of temperature (which might be used to destroy infectious bacteria selectively).[17]

Though scientific research and scholarly communication (especially
across German borders) became increasingly difficult, the young field of
biomedical engineering and biophysics continued its growth. Indeed, what
may be the world's first biophysical conference took place in Germany in
1941. The site was Oberschlema, in Saxony, near the Czech border, and
Schwan attended.[18] The leading figures were Boris Rajewsky and Nikolai
V. Timofeeff-Ressovsky (famous for his genetic studies of fruit flies through
radiologically induced mutation). The main topic was the biological effects
of ionizing radiation, but the effect and medical application of nonionizing
radiation were also discussed. Two years later the German Biophysical
Society was formally established in Berlin. (This society was dissolved at
the end of the war, but a few years later a new biophysical society was
formed in West Germany.)[19]

One day in the late fall of 1941, Rajewsky unexpectedly told Schwan that
he was ready to receive his Ph.D. and that he would have to take his math
exam that very day. Expeditiousness was called for because the examining
math professor had been drafted and was about to leave the university. The
result was that Schwan completed his Ph.D. before his research had reached a
conclusion. The research, though, continued and was eventually completed.

The winter of 1942–1943, which saw the German army's disaster at
Stalingrad, brought a marked worsening of conditions in Frankfurt. Meat,
fish, butter, and many other ordinary goods became hard to obtain, travel
became more difficult, and the bombing of the city began. Until then
Schwan was able to avoid military service because of his medical condition,
but exemptions from military service became much harder to obtain and he
would have been drafted except for the determined efforts of Rajewsky, the

Director of the Institute. (Rajewsky had joined the Nazi Party, but this was, in Schwan's view, an indication of opportunism rather than conviction; Rajewsky knew of Schwan's anti-Nazi feelings, but supported him nevertheless.)

Until 1943 Schwan was permitted to continue his research into the electrical properties of biological materials, but then, like everyone else at the Institute, he was compelled to work on military projects. At first he worked on a device to measure water droplets that was to be placed in a radiosonde (a combination of meteorological instrumentation and radio transmitter carried aloft by a balloon). This he did only part time, but in the summer of 1943 Rajewsky came back from a meeting in Berlin with word that henceforth all work must contribute to the war effort. Schwan was told to work full time on Project Chimney Sweeper, whose objective was a countermeasure to radar detection of submarines.

Shipborne and, especially, airborne radar became so effective in detecting surfaced submarines that in the summer of 1943 the German Admiral Doenitz conceded, "The method of radio location which the Allies have introduced has conquered the U-boat menace."[20] The Germans countered with a retractable air intake, called *Schnorkel*, that allowed a submarine to travel submerged, though at a shallow depth, while running on the diesel engines and recharging batteries. In January 1944 German submarines began to be equipped with *Schnorkel*.[21] The Allies responded, through Project Hawkeye, with modifications to existing radar equipment that made the exposed part of the *Schnorkel* detectable. The move-countermove continued as the Germans, through Project Chimney Sweeper, developed an antiradar covering, called *Sumpf* (swamp), that could be placed on *Schnorkel*.

Sumpf was an early version of the technology that more recently (along with other antiradar measures) is used to make the stealth bomber almost invisible to radar. It was well known that a reflecting surface could be made into an absorbing surface by coating it with a quarter-wave layer, that is, a partly transparent layer whose thickness is one-quarter the wavelength of the light or other electromagnetic radiation passing through it. By placing several layers on a surface, one could render it invisible to radar operating at several wavelengths. What Schwan was asked to do was measure the electrical properties of different materials that might be useful as coatings. For this work, involving microwaves of wavelength as short as 10 centimeters, Schwan was provided with two of the most advanced German magnetrons (built by Siemens).

Reluctant to contribute to the war effort, Schwan sought ways to delay taking the data that would be useful to Project Chimney Sweeper. The Allied bombing provided one excuse. The Institute was, in fact, heavily damaged in a series of raids in March 1944. (These raids were so destructive that a large proportion of all buildings in Frankfurt were rendered uninhabitable.) "Necessary" improvements to instrumentation provided another excuse. Finally, however, he had to report the data, but he did make one attempt to leak to the Allies the frequencies at which *Sumpf* was effective. (Schwan was so worried that this attempt would be discovered by the German authorities that he purchased a handgun with the intent to shoot himself when the Gestapo came banging on the door.)[22] The Germans did effectively employ *Sumpf* in late 1944, but too late to have much effect of the course of events.

In February 1945 US and British forces moved into Germany, and in March General Patton's army approached Frankfurt. Most of the Institute staff had long since moved to facilities more centrally located. About half a dozen people remained at the Institute in Frankfurt, and they asked Schwan to represent them, partly because he spoke English. On 15 March there were reports of approaching US forces, and at 9 P.M. the residents of Frankfurt received the order to evacuate the city. Schwan elected instead to move to the Institute—he suspected that the Americans would arrive there first—and he crossed from the north side of the river, where he lived, to the south side, where the Institute was located, at about 11 P.M. Just a few hours later all of the city's bridges over the Main were blown up by the Germans.

Because Patton's army bypassed the city, it was about 20 March when Frankfurt was finally occupied. Schwan represented the people at the Institute, and he cooperated fully with the US military. Indeed, he was one of the first Germans to receive a pass to cross the temporary bridge constructed across the river (Figure 3). (On the north side of the river was not only his apartment but also the military government of the city, where Schwan went to ask for coal and other supplies for the Institute.)

It was at this point that Schwan became involved in one of the largest transfers of scientific and technical know-how in history. This was the transfer, in the years 1945 through 1948, from Germany to the United States, the Soviet Union, Britain, and France of German scientific and technical information, equipment, and experts.[23]

In August 1944 a British-US collaboration, called the Combined Intelligence Objectives Subcommittee (CIOS), was established to identify and exploit German documents, matériel, and persons that might contribute to shortening the war.[24] CIOS, whose mission was soon extended to include the gathering of information of scientific and industrial value, sent hundreds of investigators to examine sites and interrogate scientists and technicians in liberated Europe and in Germany and Austria. Beginning in May 1945, much of the US effort to gather scientific and technical information in Europe was carried out by the Field Information Agency, Technical (FIAT) of the Office of Military Government for Germany.[25] FIAT undertook an ambitious program to microfilm documents and to publish reviews of the information available.[26]

Figure 3. Just ten days after Patton's forces moved into Frankfurt, Schwan received this pass from the US authorities to cross the Main over the hastily constructed temporary bridge.

For more than two years, from the first days of the US occupation, investigators visited the Institute for the Physical Foundations of Medicine and asked Schwan and others about the work there. Of foremost interest was the Institute's relatively minor involvement in the German atomic pile project; the Institute was to have studied the effects of intense radiation. A powerful electron accelerator (a type of cascade generator consisting of three transformers on top of each other) had been designed to accelerate a 10 or 100 milliamp current to three million volts to produce an intense x-ray beam. The accelerator was never completed. The tower that was to house it was destroyed in a small-scale bombing raid (which suggests to Schwan that it was targeted). The transformers were placed in a cave in what became the Soviet Zone of Germany and were later taken to the Soviet Union.

Schwan was questioned about the antiradar covering and his other work at the Institute. He told the first investigators that many of his instruments had been transported about a hundred miles northeast of Frankfurt. Though the site was then just behind the line of Allied advance, Schwan went with two US soldiers in a jeep to locate his equipment. The soldiers recovered his high-frequency equipment, including a variable-frequency generator and two special magnetrons used to drive coaxial lines, but allowed him to keep only his 1-gigahertz equipment. Other investigators who came to the Institute microfilmed a large number of documents and asked for reports. Schwan was asked to write up his own research results, and he contributed six FIAT reports, all published in 1947 and 1948.[27]

Since Rajewsky had been a Nazi Party member, he could not serve as Director of the Institute (see dialogue below), and soon after V-E Day he asked Schwan to be acting director. Schwan, who had already proved his theoretical and experimental abilities, soon showed that he possessed adminstrative abilities as well.[28] In addition to attending to the many needs of the Institute, he was able, within weeks of war's end, to get some of his equipment to work and gather data on the conductivity and dielectric constant of blood in the range 100 to 1000 megahertz.[29]

SCHWAN: *... The next comment pertains to loyalty and disloyalty in general. How far should you go with your responsibilities in this regard? Well, I chose to try to transmit information to the outside. I didn't tell you about another experience I had which had a negative effect on me. After the war, Rajewsky approached me and asked me to sign a statement that he was always anti-Nazi, in spite of the fact that he had been a member of the Nazi Party. He wanted me to attest that he saved a number of Jews from persecution, hiding some of them in his house.*

NEBEKER: *Were those statements true?*

SCHWAN: *I didn't know. I was initially weak enough to sign the statement. I had considerable qualms about signing it. A couple of days later I came back to him and said, "Herr Professor, I've written another statement in which I stated that although you knew I was an anti-Nazi, you gave me a job." I believe that was the case indeed. I said, "I shouldn't have signed the other statement. I don't know anything about your hiding Jewish people and so on." He gave me back the original certificate and accepted the weaker, second statement. But I know he considered me disloyal.... In retrospect today, I don't know if I would take those two actions again. I'm convinced that nine out of ten people—if not 99 out of 100 people—wouldn't act as I had acted on similar occasions. It's very, very demanding, of course. While Rajewsky frequently recommended me for all sorts of things like the Rajewsky Prize and other things, he nevertheless felt that I couldn't be trusted. His wife told me later that he admired me, but that he felt that I was not trustworthy. So much about loyalty and one's responsibilities to mankind and one's society.*[30]

In 1946 Schwan was awarded his *Habilitation*, which is a degree beyond the Ph.D. that in Germany qualifies one to teach at a university. His thesis, entitled "The determination of the dielectric properties of semiconductors, especially biological substances in the decimeter wave range," dealt largely with the techniques of measurement and drew on Schwan's improvements in instrumentation for Project Chimney Sweeper. (Some of the materials he investigated for that project were similar to biological materials in that they were highly conductive and with large dielectric constants.) One of the FIAT reports mentioned above was, in fact, this *Habilitationsschrift*.

One of the US investigators was a Navy lieutenant by the name of David Goldman. He had received one of the first Ph.D.s in biophysics in the United States (from Kenneth S. Cole at Columbia University), and had already contributed to the field by analyzing ion conductance across a membrane by means of what is today called Goldman theory.[31] Goldman visited the Institute several times and talked at length with Schwan about his work on the electromagnetic properties of biological materials. It may have been Goldman's initiative that led to Schwan's emigration to the United States.

In addition to equipment and information, US authorities—like Soviet, British, and French authorities—sought German and Austrian scientists and engineers. Project Overcast, which from the summer of 1945 to the summer of 1946 brought to the United States some 160 specialists in rockets and jet aircraft, evolved into a broader program, called Project Paperclip, to attract civilian as well as military experts.[32] In Schwan's words:

One day a lieutenant came to me and said, "Here is a contract. Think it over. I won't put any pressure on you. I'll come back after a week, and you can tell me if you'd like to accept it or not." The contract specified that I would be transported for half a year to the United States and back. Quarters in the United States and aboard ship would be that of a junior officer. I would be on leave from the university. My salary would continue to be paid in Germany at double the rate. In the United States I would get free lodging and free food in the officers' canteen, and a per diem of six dollars a day. That was almost too good to believe, and so I accepted.[33]

Schwan was eager to learn about work done in the United States and elsewhere: "Germany had been isolated scientifically since the Nazis came to power. Here I had a chance to catch up in knowledge again, and hopefully do some good work."[34] He was assigned to the Aeromedical Equipment Laboratory at the Navy base in Philadelphia, and he arrived in early September 1947. One of his first tasks was to study the noise spectrum of jet engines (from ultrasonic through audible to very low frequency vibrations) and to determine the bioeffects of such vibrations (for which purpose he obtained appropriate sonic equipment). An immediate application of this work was to design "ear defenders" to protect human hearing against engine noise.

A Research Program in the United States

Schwan expected to work only six months in the United States and to work only on assigned projects. It therefore surprised him when, shortly after arriving at the Aeromedical Equipment Laboratory, he was invited to submit a proposal for research he would like to do. He proposed an ambitious, long-term investigation of the electrical properties of biological materials, using state-of-the-art instrumentation to extend the investigation to frequencies higher and lower than those studied earlier. The Navy approved his proposal, and in 1948 Schwan began the research program that was to occupy him for most of the next four decades.

This was a new, fundamental approach to biophysics. Almost all previous work had been directed toward particular questions raised by physicians or biologists, such as what is the electrical behavior of a neuron or what are the effects of x-rays. Schwan proposed focusing on the biological materials themselves, determining the full range of their physical properties, including how energy in various forms interacts with the biological materials. The resulting understanding of the electrical properties of membranes, cells, and tissues could later, he believed, serve as a basis for solving problems encountered in research, diagnosis, and treatment. As Schwan put it, "What a physicist does, if he becomes interested in some material, is to measure its properties without any question about how he can apply the information. As he measures the properties, he asks himself, "Why are the properties as observed?" To this he adds the question, "How does energy interact with the material?" If he knows why the properties are as observed, he can explain how energies interact with it. Finally, out of this comes intelligently undertaken application."[35]

What made Schwan's proposal especially attractive was his mastery of the principles of the instrumentation that promised to yield new knowledge. One of his first accomplishments was building a bridge circuit for the accurate determination of impedance of highly lossy substances at low frequency.[36] (See Figure 4.) The reactance of most biological materials, such as intact tissues or cell suspensions, was impossible to measure at low frequencies with existing instruments because it is much smaller than the resistance. One needed to build an instrument of high resolution and—the greatest challenge—to calibrate the small reactances of that instrument in the frequency range of interest. Schwan located a manufacturer (Leeds and Northrup) willing to build a variable conductance box to his specifications, and he designed a procedure for achieving the high-precision calibration over the entire range of the bridge circuit.

The building, calibrating, and testing of the instrument took more than a year, but its use quickly yielded important results. Schwan began using it to measure the capacitance of muscle tissue in the frequency range 10 hertz to 100 kilohertz. The measurements at higher frequencies confirmed some earlier results, while those at lower frequencies, which no one earlier had succeeded in taking, revealed an unexpected capacitative dispersion (that is, a step-like change in capacitance).[37] Schwan was able to account for this capacitative dispersion as the result of a hitherto unsuspected mechanism of dielectric relaxation in muscle tissue.

Figure 4.   This photograph, from about 1955, shows Schwan with an impedance bridge.

In his investigation of the electrical properties of biological materials, Schwan focused on two fundamental properties, conductivity and permittivity. (Conductivity is defined to be the ratio of induced current to applied field and equals the reciprocal of the resistance per unit volume; permittivity is related in a simple way both to capacitance and to the dielectric constant, which is the ratio of the strength of an electric field in vacuum to that in the material for the same distribution of charge.) A knowledge of these two properties allows one to calculate many other properties of the material. By Mie theory, for example, one could calculate how electromagnetic energy is in part scattered by and in part propagated through a living organism. Schwan had begun this work in Frankfurt; now vastly superior instruments were available.

The first step in Schwan's program was the accurate determination of the electric properties of biological materials. Before the war, researchers had measured the dielectric constant of biological materials for frequencies from 1000 hertz to 10 megahertz. Schwan, in a study of muscle tissue, extended the range from a few hertz to 1 gigahertz—almost nine decades as compared with the four decades achieved earlier.[38] (Additional data, communicated to Schwan by Julia F. Herrick and others at the Mayo Clinic, extended the range to 8.5 gigahertz.) Measurement at the low frequencies required the high-resolution bridge mentioned above. Measurement at the high frequencies was based on controlled propagation of electromagnetic waves and used transmission line and waveguide components. The resulting graph (Figure 5) shows that the dielectric constant changes in distinct stages, called dispersions, which were labeled α, β, and γ.

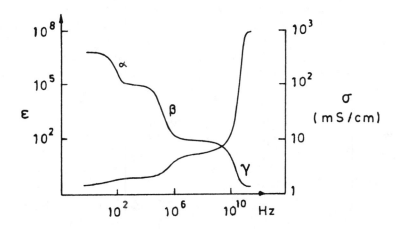

Figure 5. This graph of the dielectric constant of muscle tissue as a function of frequency shows the three dispersions, labeled α, β, and γ.

The second step of Schwan's program was to explain the measurements in terms of the atomic, molecular, and macromolecular structures of the material, that is, to give biophysical explanations of the properties. Two pioneering biophysicists, Kenneth Cole and Hugo Fricke, had earlier discovered and explained the β dispersion as resulting from electrical charging of cell membranes. Schwan identified the γ dispersion as that of water and found that the α dispersion was the result of a variety of processes, including the slow charging of intracellular membranes and the polarization of the counter-ion cloud that surrounds the surface of charged membranes.[39] He used a variety of methods to elucidate these mechanisms, especially bulk measurements on different preparations of tissues, cells, and cell components.

The decision to investigate the electrical properties of cell suspensions and tissues at low frequencies was in part motivated by the work of Fricke and Cole at higher frequencies. Both had postulated that the cell membrane properties displayed "constant phase angle" behavior similar to that observed at the interface between a metal electrode and electrolyte. It is characterized by a frequency dependence of both resistance and capacitance such that the electrical phase angle of the interface does not change with frequency. But I thought that the behavior at higher frequencies could be explained by different postulates. Variability in cell size and shape and its interior components could simulate observed deviations from single time constant behavior. Such behavior is predicted by an equivalent circuit which puts the membrane capacitance in series with internal and extracellular fluid conductivities. Such single time constant behavior could be anticipated only for uniform cell size and spherical shape and no internal content of organelles and proteins. I also knew that red cell suspensions approximate the single time constant behavior much better than tissues since they better fulfill at least some of the necessary assumptions. It became apparent to me that extension to lower frequencies would more clearly show if membranes display the constant phase angle law or not. However, it also became clear that this would require equipment of high resolution, able to detect with accuracy the small capacitive current component which reflects the membrane capacity. Conductance determinations alone would be inconclusive since the low frequency conductance is dominated by the strong contribution of the extracellular fluids. The question how the membrane behaves electrically was to emerge as a major topic of biophysical interest. To answer the question raised by the constant phase angle model was therefore important. Today, this constant phase angle concept is no longer used, and the conductive properties of the membrane are linked to its channels as first formulated by the famous Hodgkin-Huxley model.[40]

In thus relating electrical properties to biological structure, Schwan was contributing to a research tradition going back to the nineteenth century. An outstanding example was work done about 1910 by the German physiologist R. Höber, who showed by measurements of conductivity of red blood cells at different frequencies that such a cell consisted of an electrically conducting interior surrounded by a membrane of high resistance.[41] At that time the physical structure of a living cell was largely unknown, and Höber's argument was hotly contested but soon confirmed by other sorts of investigation.

It happened that Höber had left Germany in 1936—his wife was Jewish—and in 1948 was a research professor at the University of Pennsylvania, situated not more than a few miles from the Aeromedical Equipment Laboratory. He knew of Schwan's 1948 paper, and when he learned that Schwan was in town he invited him to give a seminar for the Philadelphia Physiological Society and the Department of Physiology of the University of Pennsylvania's medical school. This led first to Schwan's serving as consultant to several departments at the School of Medicine and then, in 1950, to Schwan's appointment as assistant professor in the Department of Physical Medicine.

Schwan welcomed the move to the university for several reasons. A number of people there were interested in the sort of work he was doing. Besides Höber, there was the cardiologist Calvin Kay, who was working to improve electrocardiographic techniques, and there was an electromedical laboratory at the Moore School of Electrical Engineering.[42] Schwan also valued an academic setting, with the opportunity to teach and to direct the research of graduate students.

Because the Navy permitted him to take the equipment he had assembled and built at the Aeromedical Equipment Laboratory, Schwan quickly established a laboratory at the university. It was not long before he had two Ph.D. students: Ed Carstensen, who was interested in the acoustic properties of biological materials, and Kam Li, who worked on microwave properties. In 1952 Schwan was named associate professor in the School of Medicine and was also appointed head of the Electromedical Group (renamed the Electromedical Division in 1954) at the University of Pennsylvania's Moore School of Electrical Engineering.

Schwan was familiar with studies of ultrasound from his days at the Frankfurt Institute, especially the work of Justus Lehmann. He believed that the acoustic properties of biological materials could be profitably investigated with an approach similar to the one he was already taking to investigate electromagnetic properties: investigation of frequency dependence and the step-by-step identification of which tissue and cellular components are responsible for which phenomena. Moreover, there were striking similarities in the mathematical descriptions and some similarities in the required instrumentation. Thus he was happy to add bioacoustics to his research program to accommodate the graduate student Ed Carstensen, who came to the University of Pennsylvania with prior training and a strong interest in bioacoustics.

Schwan and Carstensen studied the acoustic properties of blood and discovered the mechanism of ultrasonic absorption. They found that, unlike the absorption of electromagnetic waves of low and radio frequencies, where the cell membranes are of paramount importance, ultrasonic absorption occurs mainly by the proteins within the cells. Some years later Schwan collaborated with another graduate student, Helmuth Pauly, on acoustic properties of tissues. The caliber of this work is indicated by the selection of four of Schwan's papers for inclusion in the *Benchmark Papers in Acoustics* volume on ultrasonic biophysics.[43]

The third step of Schwan's research program—after measuring physical properties and learning what structures and mechanisms account for them—was to use this understanding in applications. As Schwan put it, "Don't just observe properties. You must understand what's going on, and then you gain predictive power."[44] Schwan and his coworkers were involved in many areas of application. Most significant were his contributions to evaluating health hazards of electric fields, and this is dealt with at some length below. Also very significant were extensive studies of the linear and nonlinear properties of bioelectrodes.

Many of the instruments used in biomedical research and in medical diagnosis include bioelectrodes, as do devices (such as cardiac defibrillators and pacemakers) used in medical treatment. Hence, understanding electrode behavior, especially electrode artifacts and electrode polarization (resulting from the electric potential generated at the interface of the electrode and the tissue), is of widespread practical importance. Schwan and his collaborators studied electrode behavior over a wide frequency range and modeled the observed frequency dependence, and they discovered at what current levels nonlinear effects become important. Going beyond understanding the phenomena, they developed techniques for eliminating, minimizing, or compensating for the effects of electrode polarization. This work has been widely cited, and in 1992 the *Annals of Biomedical Engineering* published a special issue on electrodes in honor of Schwan.[45]

In 1948, through friends made at the International House of the University of Pennsylvania, he became acquainted with Anne Marie Del Borello. They were married the following summer, and the family grew over the next twelve years with the arrival of four daughters and one son. In 1952 Schwan became a naturalized citizen of the United States.

Through the 1950s and the following three decades, Schwan's research program flourished. Schwan and his coworkers continued work in the areas described above. They gathered more data on the electrical properties of tissues, cells, and cellular organelles. They elucidated the mechanisms responsible for certain properties, notably counter-ion relaxation (caused by the movement of ions along a cell or particle surface) and protein bound water relaxation (according to which water molecules at the surface of protein molecules have a behavior intermediate between frozen and liquid water).[46]

In 1957 Schwan published a 62-page review article entitled "Electrical properties of tissue and cell suspensions."[47] This article provided for the first time a comprehensive account of the field and soon became frequently and widely cited. Another extremely influential publication appeared in 1962, in which Schwan and coworkers showed that the α dispersion (mentioned above) occurred in nonbiological colloidal systems.[48]

Schwan conducted extensive investigations on the propagation of electromagnetic energy into tissues and into the human organism as a whole. He and coworkers determined reflection and absorption coefficients and how energy passed through complex tissue arrangements. They built a novel unechoic facility for these measurements and made use of mannequins filled with fluids simulating tissues.[49] In other work, Schwan, H. N. Kritikos, and K. R. Foster studied in detail temperature increases in different tissues (including the possibility that microwaves might generate localized "hot spots") and the effects on the human thermoregulatory system.[50] Some of this work involved lengthy calculations, and Schwan made much use of computers. It was by calculation that Schwan and coworkers discovered a resonance effect in the absorption of microwaves; they later confirmed the effect experimentally.[51]

In almost every area of his work, Schwan contributed also to the practical application of the knowledge or techniques. Besides study of the health effects of microwaves (discussed below) and the already mentioned work on bioelectrodes, there was important work on diathermy (the generation of heat in tissues for medical purposes), on impedance plethysmography (which allows determination of blood volume changes in tissues and hence is used to diagnose venous occlusions), and on electrophoresis (a research and diagnostic technique based on the different mobilities of suspended particles in electric fields). Schwan's work on suspensions of red blood cells led to his development with R. H. Okada of a hematocrit (which electronically measures the number of cells in a blood sample), that was successfully built and marketed.

Schwan and coworkers pursued applications of their investigations in bioacoustics, just as they did in their study of ultrasonic diathermy, where they compared its effectiveness to that of radio-frequency and microwave diathermy. One of the most important applications of bioacoustics to date, echocardiography (a procedure for recording cardiac structure and function using ultrasound), was developed at Schwan's laboratory. In 1957 John Reid, who had already gained recognition for his pioneering work on ultrasonic visualization with John J. Wild in Minneapolis, entered the Ph.D. program at the University of Pennsylvania. Schwan secured NIH support for and supervised Reid's research, which was carried out in cooperation with cardiologists Calvin Kay and Claude Joyner of the Penn Department of Medicine.

Establishing Biomedical Engineering as a Discipline

As mentioned at the beginning of this article, biophysical research long predates World War II. But work was concentrated in two areas—electrophysiology (especially the study of excitability and contractibility and the study of electrical conductance of blood and tissues) and biological effects of ionizing radiation. There were few biophysical institutions. The Kaiser Wilhelm Institute at Frankfurt, the Johnson Foundation for Medical Physics at the University of Pennsylvania, and the Biophysics Department of the Cleveland Clinic have already been mentioned. There were a few others, including the laboratory established by Hugo Fricke at Cold Spring Harbor on Long Island, Kenneth Cole's biophysical laboratory at Columbia University, and the Siemens research laboratory at Erlangen headed by J. Paetzold. It may be that the first biophysical societies were two established during the war: the one formed by Rajewsky discussed above, and one organized in Holland at about the same time by H. C. Burger, head of a medical physics laboratory in Utrecht.[52]

During and after the war, biophysics and biomedical engineering grew rapidly, both in numbers of researchers and in range of studies. It is noteworthy that both Schrödinger's *What is Life?* (which showed how physics could elucidate biology) and the first of Otto Glasser's texts on medical physics appeared during World War II.[53] In the postwar decades there slowly emerged, in a complicated and sometimes contentious way, an institutional and organizational base for this work. From the mid-1950s on, Schwan gave a substantial portion of his time to building that base.

Schwan's great organizational achievement was the establishment of biomedical engineering at the University of Pennsylvania. Starting from a single assistant professorship (the financial support for which vanished after two years), Schwan gradually built a biomedical engineering research and teaching program that by the time of his retirement in 1983 comprised 12 primary faculty positions, an undergraduate program with about 150 enrolled, and a graduate program with about 60 enrolled.

Essential to this achievement was Schwan's success in obtaining outside funding. In 1952 he won support from the Office of Naval Research, which continued until 1978. Also in 1952 the National Institutes of Health (NIH) and the Air Force approved grant requests. Schwan's NIH grant, under the title "Electrical and Acoustic Properties of Biological Materials," was approved and extended for a total of 25 years. The Air Force support continued until 1964.[54]

In 1960 Schwan won university approval of an independent Ph.D. program in biomedical engineering and shortly thereafter received a substantial training grant from NIH. This grant, which continued for 20 years, made it possible to attract excellent graduate students and was vital to the establishment of the Biomedical Electronic Engineering Department. Schwan was chairman until 1973, which was the year the department changed its name to the Bioengineering Department and added an undergraduate program. The University increased its support of biomedical engineering as the program grew.

This department seems to have been the first department in biomedical engineering at any university, and it served as a model for programs elsewhere. Before 1970, Ph.D. programs had begun at a few other schools, including Purdue University and the University of Washington.[55] In the late 1950s Schwan met often with James Dow and others at Drexel University, also in Philadelphia. They established a master's program in biomedical engineering, which was primarily directed toward medical doctors, while the Penn doctoral program attracted mainly engineers.[56]

Schwan also played a major role in the establishment of professional organizations. Here the situation was, and remains, especially complicated, mainly because biophysics and biomedical engineering encompass a great variety of work carried out by people with very different backgrounds.

Not long after the war, the American Institute of Electrical Engineers (AIEE) formed a Committee on Electrical Techniques in Medicine and Biology, which organized its first annual conference in 1948. In 1952 the Institute of Radio Engineers (IRE) formed a Professional Group on Medical Electronics.[57] In 1953 the AIEE Committee, the IRE Professional Group, and representatives of the Instrument Society of America formed a Joint Committee for Engineering in Medicine and Biology (JCEMB). The mission of JCEMB was to organize annual conferences, and these have continued to the present.

Beginning in the early 1950s Schwan was a member of both the AIEE Committee and the IRE Professional Group, and also served on JCEMB. In 1959 and again in 1965 the annual conference was held in Philadelphia, and he acted both times as conference chair. Over the years attendance at the conferences has increased from less than a hundred in the early 1950s to nearly 2000 in the late 1980s. In January 1963 AIEE and IRE merged to form the Institute of Electrical and Electronics Engineers (IEEE). The members of the AIEE and IRE technical committees for biomedical engineering were in fact a contributing force to the merger, as almost all of them favored it and had been collaborating with their counterparts in the other society for years. Schwan was instrumental in melding the two committees into the IEEE Group on Engineering in Biology and Medicine (later the IEEE Engineering in Medicine and Biology Society).

In the mid-1950s Schwan and others considered how a biophysical society might be established. Some favored affiliation, either with the American Institute of Physics or the Federation of Biological Societies, while others thought an independent society preferable. The latter view prevailed, and in 1956 a committee, consisting of K. S. Cole, E. Pollard, Otto Schmitt, and Samuel Talbot, set to work to determine the organizational structure of what was to be called the Biophysical Society. R. Stacy was selected as conference chair and Schwan as publicity chair for the first annual conference. Schwan wrote, "I mailed more than 1000 letters, asking early recipients to let me know about others interested in a biophysical society. The response was overwhelmingly favorable."[58] Seven hundred attended the first conference, held in Columbus, Ohio, in 1957 (see Figure 6). Schwan had done so well at publicity that he was asked to continue that work, which he did for the next four years. The society grew, but assumed a biological orientation as biochemists and physiologists entered in large numbers.

The National Institutes of Health played a large role in establishing the new field.[59] Its support of the research and training at the University of Pennsylvania has already been mentioned. Schwan in large measure repaid this debt by serving for many years as member, and sometimes chairman, of various NIH study sections and councils.

NATIONAL BIOPHYSICS CONFERENCE

A steering committee of some fifty scientists, representing various aspects of biophysical research in this
country, has organized a national biophysics conference to
take place in Columbus, Ohio, from March 4 - 6, 1957. The
conference will encompass studies which employ the approach
of physics in biological measurement and theory, at levels
of organization from molecules and cells to complex systems
and psychophysics.

The program is expected to include twelve invited papers
related to different biophysical fields and a large number
of contributed papers. Scientists with biophysical interests may write to Dr. Herman P. Schwan for further details
and information on presenting contributed papers.

Samuel A. Talbot, Chairman
Program Committee
Department of Medicine
Johns Hopkins Hospital
Baltimore 5, Maryland

Herman P. Schwan, Chairman
Publicity Committee
School of Medicine
University of Pennsylvania
Philadelphia 4, Pennsylvania

Figure 6.   The 1956 announcement of the March 1957 meeting that led to the founding of the Biophysical Society.

[In the first years at the University of Pennsylvania] people didn't understand what I was doing. Cole and Fricke, yes. I met Cole and Fricke. But most of the physiologists didn't even understand Cole. I remember very well when I came to the University of Pennsylvania, that the physiologists often said to me, "Herman, you seem to understand that work of Cole. Can you explain it to us? We have no idea if that's important or not." I tried my best, of course, to do so. The physiologists and early biophysicists just were not trained enough to understand the relevance of this work. It took quite some time. I think an interest in such biophysical investigation started to develop very slowly. Interest increased after two British physiologists, [A. L. Hodgkin and A. F.] Huxley, [received] the Nobel Prize for their work on the electric properties of nerve axons. Then interest in that sort of work grew fairly fast. Cole told me that he brought my work to the attention of G. Falk and P. Fatt. My first presentation was in 1950 at the American Physiological Society meeting in Columbus, Ohio. Then I tried to publish it in physiological journals. There was no biophysical journal at that time. I was turned down twice, by two journals. I was very discouraged. And then I submitted to a German journal where it was published in '55—five years later than Columbus meeting. By '56 I did have a reputation in the field, and I was asked to write a review article on electric properties of biological materials. I wrote a long review article which was published in '57, where I reported for the first time in English about all that sort of work. There was only that German publication before and the abstract, which doesn't say much. This review article was a great success. I think it has been quoted in the Citation Index almost a thousand times. It's still being quoted since it was the first really comprehensive treatment of electrical properties of biological materials.[60]

The interdisciplinary nature of biomedical engineering made it difficult to find or to establish an organizational home for its practitioners. In the mid-1960s Schwan worked hard to bring about a change of IEEE membership rules, so that M.D.s and Ph.D.s in the biological sciences (who were not engineers) could be full members of the IEEE Group on Engineering in Biology and Medicine. What Schwan proposed was that technical groups become semiautonomous societies able to establish their own membership requirements. When this was not approved by the IEEE Board of Directors (though the designation 'group' was soon changed to 'society'), Schwan worked with three others—Jack Brown, John Jacobs, and Larry Stark—to found the Biomedical Engineering Society in 1968.[61] Schwan and others hoped that this society would bring together those trained in engineering and those trained in biology and medicine. The society, however, soon became dominated by physiologists, disappointing Schwan and never having much adverse effect on membership in the IEEE Group on Engineering in Biology and Medicine.

Schwan has been active in several other societies. He helped establish the International Federation of Medical Electronics in 1957 and was a founding member of the Bioelectromagnetics Society, which was formed in the mid-1970s. He was a member of the International

Institute of Medical Electronics and Biological Engineering and served two years on its advisory board in the late 1960s.

After his 1947 move to the United States, Schwan maintained close ties with German researchers mainly through frequent trips to Germany. In 1962 he was a visiting professor at the University of Frankfurt, and in 1986/87 at the University of Würzburg. In 1962 he received the lifetime appointment of foreign scientific member of the Max Planck Institute for Biophysics.

Although Germany was a world leader in biophysics before World War II, in the postwar period biophysics and, especially, biomedical engineering developed much more rapidly in the United States. Schwan attributes this principally to three factors: (1) the existence of many private as well as public universities fosters independent ventures, (2) the separation between undergraduate and graduate education at American universities makes it easier for a graduate program to be highly selective and to adopt an interdisciplinary position (most biomedical engineering programs, Schwan points out, began as research laboratories that added a doctoral program, and in some cases later an undergraduate program), and (3) there are many sources of outside funding, notably the National Institutes of Health.[62] There are, of course, many other reasons for American leadership in biophysics and biomedical engineering, one of which is the emigration to the United States of scientists trained in Germany, such as Max Delbrück, R. Höber, Justus Lehmann, Schwan, and Georg von Békésy.

Safety Standards for Microwaves

During World War II some servicemen expressed concern that the microwave radiation produced by radar transmitters might cause health problems. Studies at the Naval Research Laboratory and other military laboratories failed to reveal any adverse effects. In the immediate postwar years, as microwave oscillators of greater and greater power became available, more people expressed concern. In 1953 the US Navy convened a meeting on microwave health hazards. The attendees—including Kenneth Cole, David Goldman, James Hardy, and Schwan—arrived at 100 mW/cm^2 as the dividing line between safe and hazardous exposure. Because there was no evidence that harmful effects occurred except through heating, the principal consideration was whether the human body could readily dissipate the heat generated by the microwaves. After the meeting, Schwan recalculated what heat the body would absorb and how much could readily be dissipated. He decided that 100 mW/cm^2 might not be safe and proposed instead a standard of 10 mW/cm^2.[63] (Schwan thought that setting a standard that was anything but a power of 10 would give an improper impression of accurate knowledge of what the dividing line ought to be.)[64] The Navy quickly adopted Schwan's proposal.

In 1955 Julia Herrick of the Mayo Clinic organized what may have been the first scientific conference on microwave bioeffects.[65] Among the fourteen papers presented was one by Schwan and Kam Li that argued for a 10 mW/cm² standard.[66] This paper was published the following year, as was another by the same authors entitled "Hazards due to total body irradiation by radar."[67] Over the following decades the number of articles specifically on the health hazards of electromagnetic radiation steadily increased, and Schwan has been a prolific contributor to this literature.[68]

In 1959 the American Standards Association, now the American National Standards Institute (ANSI), agreed to establish a committee, which would be jointly sponsored by AIEE and the Navy's Bureau of Ships, for the purpose of setting a health standard for microwave exposure.[69] Schwan was selected as chairman of the committee, which was designated C95, and at its first meeting in February 1960 he set up six subcommittees, designated C95.I through C95.VI, each investigating a particular issue. Things moved extremely slowly, mainly because most committee members seemed to be too busy to give much time to the effort, but also because many committee members were unwilling to make recommendations on the basis of available evidence. Schwan repeatedly emphasized that the committee's task was not to undertake research, but to decide on the basis of what was already known.

Particularly dilatory was Subcommittee IV, charged with setting a safety level for human exposure to microwaves. After two years, during which time the subcommittee accomplished almost nothing, Schwan reconstituted C95.IV and assumed the chairmanship himself. The evidence available to the subcommittee included (1) physiological knowledge about metabolic rates and tolerances to heat, (2) work on mode of propagation of electromagnetic radiation in biological materials and on dosimetry of energies absorbed, (3) diathermy experience (which involved microwave radiation at levels far above suggested standards), and (4) the literature specifically on the possible health hazards of microwaves.[70] C95.IV recommended the standard of 10 mW/cm², and this was approved by the whole committee in November 1966 as ANSI Standard C95.1-1966.

This standard aroused some dissension within the committee and some controversy outside. There seemed to be, at that time, more concern that the standard was too strict than that it was too lax.[71] Some critics contended that Schwan had had too much influence in the setting of the standard. Because of this criticism (and also because he found himself doing too much committee work, especially for IEEE and NIH), Schwan decided in 1965 to resign the chairmanship of C95. Thus, the periodic review and modification of the standard (required by ANSI rules) would be done entirely by others. Schwan felt vindicated when the first review, completed in 1974, simply confirmed the earlier standard.[72]

In 1977 a widespread and intense public interest in possible hazards of microwave radiation appeared rather suddenly. Defective television sets and leaking microwave ovens had attracted considerable attention beginning in the late 1960s, and in February 1976 the State Department made public much of its information about the microwave bombardment of the US Embassy in Moscow. (In much of the period from the early 1950s into the 1970s, the Soviets directed low-intensity microwave radiation at the Embassy building.) Probably most important in arousing public interest were the efforts of the science writer Paul Brodeur.[73] He expanded two articles that appeared in *New Yorker* (December 1976) into a book, published in 1977 and entitled *The Zapping of America: Microwaves, Their Deadly Risk, and the Cover-Up.*[74]

Schwan was drawn into the public debate. He had already given many hours of testimony before governmental bodies: the Senate Committee on Commerce in 1967 and 1968, the Congressional Committee on Science and Astronautics in 1973, and the New York State Public Service Commission from 1976 to 1978.[75] Schwan had always argued that additional research was called for, especially to determine if there were genetic or cumulative effects of microwave radiation.[76] Accordingly, he thought that safety standards would gradually change to reflect increased understanding of the effects of microwaves. Yet Schwan repeatedly found himself in the position of defending the 10 mW/cm^2 standard against those who believed the limit should be set much lower.

Some critics have charged that Schwan was unaware of Soviet research indicating "weak effects" of microwaves. In fact, Schwan met on many occasions with the leading Russian researcher Z. V. Gordon and discussed her work in detail. Other critics have claimed that Schwan was considering only thermal effects of microwaves. This too is mistaken. A principal point of contention at the Frankfurt Institute in the late 1930s was the reality of nonthermal effects.[77] Some argued, erroneously as it turned out, that the pearl chain effect, which was undoubtedly a nonthermal effect, occurred at weak fields. There were other reports of nonthermal effects that were unfounded. As a result of this experience and his understanding of the relevant biophysics, Schwan was thereafter quite skeptical of alleged nonthermal effects at weak fields.[78]

Schwan readily acknowledges the difficulties of setting safety standards for electromagnetic radiation and for electric and magnetic fields: The range of possible effects is large, deciding how to measure exposure is problematic (biological effect may be highly dependent upon what biological structures are exposed, upon frequency, upon whether the fields are intermittent, or upon how the fields are modulated), epidemiological evidence is usually ambiguous, and cumulative and long-term effects are especially difficult to determine. These questions are very much open, and the amount of research aimed at answering them has continued to increase (costing today about $25 million annually).[79]

A Diverse Career

Throughout his career, research has been the center of Schwan's activities. He has published more than 250 papers in journals of engineering, medicine, biology, and physics. In recognition of his research contributions, he has received numerous awards, including the Boris Rajewsky Prize for Biophysics, the Humboldt Prize of the Humboldt Foundation, and the W. J. Morlock Award of the IEEE. He was the first recipient of the d'Arsonval Medal of the Bioelectromagnetics Society. In 1982 the journal *Bioelec-tromagnetics* published a festschrift in his honor, and in 1986 the National Council on Radiation Protection and Measurements selected him as the Tenth Lauriston S. Taylor Lecturer. He has been named Member of the National Academy of Engineering, Foreign Scientific Member of the Max Planck Society, and Honorary Member of the German Biophysical Society. In 1983 he received the IEEE Edison Medal "for a career of creative endeavor by which engineering, physics, biology, and medicine have been amalgamated into a coherent field of electromagnetic bioengineering."

[Vladimir Zworykin] became very much interested in telemetry. . . . In the RCA laboratories he developed the technique of a pill that serves as a monitor. A person would swallow it, and it would go down through the esophagus and stomach and so on. You then could observe all sorts of very useful information, including pH and temperature. The "pill" could send them out to the receivers while travelling through the body.

. . . When Zworykin was vice president of RCA, he was the prime mover in the development of the first American electron microscope.

. . . We met fairly frequently. We had quite a discussion about certain regulations and the constitution of the International Federation [of Medical Electronics], which were adopted in due time. There was always the problem of proper representation of countries. I was primarily responsible for the International Federation adopting what I called the Logarithmic Rule. Not every country has the same vote. I suggested a logarithmic rule. Small countries, say up to 10 members, had one vote. Countries with up to ten times more members had two votes, and countries with ten times more, three votes. That worked out satisfactorily.

. . . His mind was always active with regard to all sorts of things. . . . For example, he thought of a little pocket calculator that you can carry with you, which has your total medical information on it. He envisioned a phone hook-up, sort of like a modem-like hook-up, where whenever you have a symptom, you could get properly hooked up to a specialist . . . they listen to your heartbeat and other relevant things.[80]

As the preceding account makes clear, Schwan has had many other activities. As a teacher he helped establish graduate and undergraduate programs in biomedical engineering, introduced many new courses, and supervised about twenty Ph.D. theses and about ten Master's theses (see Figure 7). According to Schwan, about half of his Ph.D. students became head of bioengineering departments or programs that they helped establish.[81] He received the American Society of Engineering Education Award in 1983, and in 1986 the University of Pennsylvania presented him with an honorary doctorate. He has been a peripatetic lecturer, giving almost 400 conference presentations and invited lectures in the period since the early 1950s. And he has served his profession in other ways: as advisor to government (to NIH and NSF, to the Department of Health, Education, and Welfare, to the Navy and the Army, to the Veterans Administration), as consultant to industry (to Bell Laboratories, General Electric, and many other companies), as organizer of conferences and conference sessions, and as journal editor and reviewer.

In recent years Schwan has written extensively on the history of biomedical engineering: on the gradual expansion of research topics and on the development of particular research lines,[82] on the evolution of health standards for radio frequency and microwave radiation,[83] and especially on institutional history.[84]

Figure 7. This 1990 photo shows Schwan with eleven of his former students. Those pictured are, from upper left to lower right, Dov Jaron, Richard Beard, Willis Tompkins, Dennis Silage, Lee L. Huntsman, Robert E. Yantorno, John Li, John M. Reid, Banu Onaral, Schwan, Zenka Delalic, and Maryam Moussavi. Among Schwan's other former students are Edwin L. Carstensen, Clifford Ferris, and David B. Geselowitz.

Schwan has himself lived through most of the history of his discipline. In late 1937, when he began work at the Kaiser Wilhelm Institute for the Physical Foundations of Medicine, there were probably only a few dozen people doing what today would be called biophysics or biomedical engineering, and most of the research concerned electrophysiology or the effects of ionizing radiation. Today there are tens of thousands of biophysicists and biomedical engineers, and their research spans an enormous range. There are a dozen or so professional organizations for these scientists and engineers, and several dozen universities have programs leading to advanced degrees in biophysics and biomedical engineering.

Herman Schwan contributed to this growth in several ways. He pioneered new research areas: dielectric properties of biological materials—from molecules to whole organisms—at high and low frequencies, the propagation of electromagnetic energy in biological materials, and the ultrasonic properties of biological materials. He achieved both accurate measurement of properties and explanation of many of the observed values. Furthermore, he applied the resulting biophysical understanding to practical problems: understanding electrode effects, developing new diagnostic and therapeutic instruments, and helping to set microwave safety standards. And he helped build the institutional basis—both at the University of Pennsylvania and in several thriving professional organizations—for the new discipline.

---

[1] L. A. Geddes, "Clinical engineering and the background of interdisciplinary engineering," *Medical Instrumentation*, vol. 9, 1975, pp. 239–249.

[2] Kenneth S. Cole, *Membranes, Ions and Impulses: A Chapter of Classical Biophysics* (Berkeley: University of California Press), pp. 1, 67–68.

[3] W. E. Röntgen, *Eine neue Art von Strahlen* (Würzburg, 1895); H. S. Gasser and J. A. Erlanger, "A study of the action currents of nerve with the cathode ray oscillograph"; *American Journal of Physiology*, vol. 62, 1922, pp. 496–524; E. D. Adrian and D. W. Bronk, "The discharge of impulses in motor nerve fibres. Part I. Impulses in single fibres of the phrenic nerve," *Journal of Physiology*, vol. 66, 1928, pp. 81–101; L. Marton, "Electron microscopy of biological objects," *Nature*, vol. 133, 1934, p. 911.

[4] An overview is provided by L. A. Geddes in "The beginnings of electromedicine," *IEEE Engineering in Medicine and Biology Magazine*, December 1984, pp. 8–23.

[5] Margaret Rowbottom and Charles Susskind, *Electricity and Medicine: History of their Interaction* (San Francisco: San Francisco Press, 1984), p. 246.

[6] H. P. Schwan, "Early history of bioelectromagnetics," *Bioelectromagnetics Journal*, vol. 13, 1992, pp. 453–467.

[7] H. P. Schwan, "Entwicklung der Biophysik und das Frankfurter Institut für Biophysik," draft manuscript of an article to appear in the proceedings of a conference held at Schlema, Germany, September 1991.

[8] The information about Herman Schwan contained in this article comes mainly from the following sources: (1) an extensive oral-history interview of Schwan conducted by the author 26 June and 1 July 1992 (from which an edited transcript has been prepared); (2) personal communications, both by mail and telephone, with Schwan; (3) Schwan's published writings; (4) other published writings; and (5) the Project Paperclip file on Schwan in the National Archives. (The transcript of the extensive interview, letters from Schwan, a full list of Schwan's publications, and copies of many documents from the Project Paperclip file on Schwan are available at the IEEE Center for the History of Electrical Engineering.)

[9] Wilhelm Schwan, *Elementare Geometrie* (Leipzig: Akademische Verlagsgesellschaft, 1929); Gerhard Hessenberg, *Grundlager der Geometrie*, edited by Wilhelm Schwan (Berlin: Walter de Gruyter, 1930).

[10] The political impotence of intellectuals was a striking feature of German society from the late nineteenth century through World War II and became especially marked in the Third Reich (see Chapter 20 of Richard Grunberger's *A Social History of the Third Reich* (New York: Penguin Books, 1974).

[11] Interview 1992, p. 8.

[12] Interview 1992, p. 13.

[13] Interview 1992, p. 15.

[14] Interview 1992, pp. 17–18.

[15] Interview 1992, pp. 16–17.

[16] Interview 1992, p. 18.

[17] H. Schaefer and H. P. Schwan, "The question of selective heating of small particles in the ultrashortwave condensor fields," *Annalen der Physik*, vol. 43, 1943, pp. 99–135.

[18] In September 1991 a conference was held at the same site to celebrate the fiftieth anniversary of the German biophysical society. Schwan was the only attendee who had also been at the 1941 meeting. His paper, "Entwicklung der Biophysik und das Frankfurter Institut für Biophysik" (cited above), will appear in the conference proceedings. The 1941 meeting is also briefly discussed in Schwan's "Early history of bioelectromagnetics."

[19] Schwan, "Entwicklung . . . ," pp. 5–6.

[20] Quoted on p. 723 of Henry E. Guerlac's *Radar in World War II* (Tomash Publishers and American Institute of Physics, 1987).

[21] Guerlac, *Radar in World War II*, pp. 727–729.

[22] Deposition made by Hermann Muth 8 July 1947 and deposition made by Hans Holzamer 9 July 1947, both in the Project Paperclip personal file on Schwan; and Interview 1992, pp. 42–43.

[23] John Gimbel, *Science, Technology, and Reparations: Exploitation and Plunder in Postwar Germany* (Stanford CA: Stanford University Press, 1990); and John Gimbel, "Project Paperclip: German scientists, American policy, and the Cold War," *Diplomatic History*, vol. 14, 1990, pp. 343–365.

[24] Gimbel, *Science, Technology, and Reparations*, pp. 3–6.

[25] Ibid., p. 29.

[26] Ibid., pp. 60–74.

[27] Two reports appeared in 1947, as FIAT Report 1097 and FIAT Report

1099. The other four reports appeared the following year in *FIAT Review of German Science, Biophysics II.*

[28] Deposition made by Boris Rajewsky 8 July 1947, in the Project Paperclip personal file on Schwan.

[29] B. Rajewsky and H. Schwan, "The dielectric constant and conductivity of blood at ultrahigh frequencies," *Naturwissenschaften*, vol. 35, 1948, p. 315.

[30] Interview 1992, pp. 106–107.

[31] D. E. Goldman, "Potential, impedance, and rectification in membranes," *Journal of General Physiology*, vol. 27, 1943, pp. 37–60. See also Cole, *Membranes, Ions and Impulses*, pp. 196–197, and 267–268.

[32] Gimbel, *Science, Technology, and Reparations*, pp. 37–59, and Gimbel, "Project Paperclip. . . . "

[33] Interview 1992, pp. 55–56, slightly edited.

[34] Ibid.

[35] Interview 1992, p. 128.

[36] H. P. Schwan and K. Sittel, "Wheatstone bridge for admittance determinations of highly conducting materials at low frequencies," *Transactions of the AIEE (Comm. and Elec.)*, May 1953, pp. 114–122.

[37] The discovery was presented first in abstract form in *American Journal of Physiology*, vol. 163, 1950, p. 748, then in detail in "Electrical properties of muscle tissue at low frequencies," *Zeitschrift für Naturforschung*, vol. 9b, 1954, p. 245. See also Cole, *Membranes, Ions and Impulses*, p. 46.

[38] Schwan, "Electrical properties of tissue and cell suspensions," *Advances in Biological and Medical Physics*, vol. 5, 1957, pp. 147–208.

[39] For a summary of this work, see Schwan, "Electrical properties of cells: principles, some recent results, and some unresolved problems," in W. S. Adelman and D. Goldman, eds., *The Biophysical Approach to Excitable Systems* (New York: Plenum Press, 1981), pp. 3–24; and Schwan, *Biological Effects of Non-ionizing Radiations: Cellular Properties and Interactions*, Lauriston S. Taylor Lectures in Radiation Protection and Measurements, Lecture No. 10 (National Council on Radiation Protection and Measurements, Bethesda MD, 1987).

[40] Interview 1992, Schwan's footnote to p. 112.

[41] Cole, *Membranes, Ions and Impulses*, pp. 6–7.

[42] Schwan, "Biomedical engineering at the University of Pennsylvania," *IEEE Engineering in Medicine and Biology Magazine*, vol. 10, no. 3, 1991, pp. 47–49.

[43] "Determination of the acoustic properties of blood and its components" (with E. L. Carstensen and K. Li), "Absorption of sound arising from the presence of intact cells in blood" (with E. L. Carstensen), "Acoustic properties of hemoglobin solutions" (with E. L. Carstensen), and "Mechanism of absorption of ultrasound in liver tissue" (with H. Pauly), all reprinted in *Ultrasonic Biophysics*, edited by Floyd Dunn and William D. O'Brien, Jr. (Stroudsburg PA: Dowden, Hutchinson & Ross, 1976), vol. 7 of the *Benchmark Papers in Acoustics* book series.

[44] Interview 1992, p. 133.

[45] *Annals of Biomedical Engineering*, vol. 20, no. 3, 1992. Schwan's attention

to electrode properties stimulated a number of other researchers, including L. A. Geddes, to contribute to this area.

[46] H. P. Schwan and J. Maczuk, "Electrical relaxation phenomena of biological cells and colloidal particles at low frequencies," *Proceedings of the First National Biophysics Conference* (Yale University Press, 1959), pp. 348–355; and Schwan, "Electrical properties of bound water," *Annals of the New York Academy of Sciences*, vol. 125, 1965, pp. 344–354.

[47] *Advances in Biological and Medical Physics*, vol. 5, 1957, pp. 147–209.

[48] H. P. Schwan, G. Schwarz, J. Maczuk, and H. Pauly, "On the low-frequency dielectric dispersion of colloidal particles in electrolyte solution," *Journal of Physical Chemistry*, vol. 66, 1962, pp. 2626–2635.

[49] A. Anne, O. M. Salati, and H. P. Schwan, "Relative microwave absorption cross sections of biological significance," *Biological Effects of Microwave Radiation*, vol. 1, 1961, pp. 153–176.

[50] H. N. Kritikos, K. R. Foster, and H. P. Schwan, "Temperature profiles in spheres due to electromagnetic heating," *Journal of Microwave Power*, vol. 16, 1981, pp. 327–344; and W. I. Way, H. Kritikos, and H. P. Schwan, "Thermoregulatory physiologic responses in the human body exposed to microwave radiation," *Bioelectromagnetics*, vol. 2, 1981, pp. 341–356.

[51] A. Anne, O. M. Salati, and H. P. Schwan, "Microwave absorption cross section studies of man with multilayer phantoms," *Digest of the 15th Annual Conference on Engineering in Medicine and Biology*, 1962, p. 296; and O. M. Salati, A. Anne, and H. P. Schwan, "Radio frequency radiation hazards," *Electronic Industries*, November 1962.

[52] Schwan, "Organizational development of biomedical engineering," *IEEE Engineering in Medicine and Biology Magazine*, vol. 10, no. 3, 1991, pp. 25–29, 33.

[53] Erwin Schrödinger, *What is Life? The Physical Aspect of the Living Cell* (Cambridge UK: Cambridge University Press, 1944); Otto Glasser, *Medical Physics* (Chicago, 1944).

[54] Schwan, "Biomedical engineering at the University of Pennsylvania."

[55] L. A. Geddes, *A Century of Progress: The History of Electical Engineering at Purdue (1888–1988)* (West Lafayette IN: School of Electrical Engineering, Purdue University, 1988), pp. 97–102, 165–203; Arthur W. Guy, "The Bioelectromagnetics Research Laboratory, University of Washington: reflections on twenty-five years of research," *Bioelectromagnetics*, vol. 9, 1988, pp. 113–128, and L. L. Huntsman, "Biomedical engineering at the University of Washington," *IEEE Engineering in Medicine and Biology Magazine*, vol. 10, no. 3, 1991, pp. 42–43.

[56] Schwan, "Biomedical engineering at the University of Pennsylvania," and H. H. Sun, "Biomedical engineering at Drexel University," *IEEE Engineering in Medicine and Biology Magazine*, vol. 10, no. 3, 1991, pp. 44–46.

[57] L. H. Montgomery, "The origin of the professional group on medical electronics," *Proceedings of the IRE*, vol. 47, 1959, pp. 1993–1999.

[58] Schwan, "Organizational development. . . ."

[59] See J. H. U. Brown's "personal view of biomedical engineering," *IEEE Engineering in Medicine and Biology Magazine*, vol. 10, no. 3, 1991, pp. 34–37, 46.

[60] Interview 1992, pp. 114–115.

[61] Interview 1992, p. 158, and Brown, "A personal view of biomedical engineering."

[62] Schwan, "Entwicklung . . . " and Interview 1992, pp. 199–200.

[63] Nicholas H. Steneck et al., "The origins of U.S. safety standards for microwave radiation," *Science*, vol. 208, 1980, pp. 1230–1237, and Schwan, "Early history of bioelectromagnetics."

[64] Telephone interview of Schwan, 30 October 1992.

[65] Schwan, "Early history of bioelectromagnetics."

[66] Schwan and K. Li, "The mechanism of absorption of ultrahigh frequency electromagnetic energy in tissues, as related to the problem of tolerance dosage," *IRE Transactions on Medical Electronics*, vol. PGME-4, 1956, pp. 45–49.

[67] *Proceedings of the IRE*, vol. 44, 1956, pp. 1572–1581.

[68] His list of publications include some forty on the health hazards of electromagnetic radiation. Besides the ones already cited, among the most important are the following: "Absorption of electromagnetic energy in body tissues—review and critical analysis" (with G. M. Piersol) in two parts, *American Journal of Physical Medicine*, vol. 33, 1954, pp. 371–404, and vol. 34, 1955, pp. 425–448; "Relative microwave absorption cross sections of biological significance" (with A. Anne, M. Saito, and O. M. Salati), *Biological Effects of Microwave Radiation*, vol. 1, 1961, pp. 153–176; "Electrostatic field induced forces and their biological implications" (with L. D. Sher), in *Dielectrophoretic and Electrophoretic Deposition*, edited by H. A. Pohl and W. F. Pickard (New York: Electrochemical Society, 1969); "Microwave radiation: biophysical considerations and standards criteria," *IEEE Transactions on Biomedical Engineering*, vol. BME-19, 1972, pp. 304–312; "Nonthermal cellular effects of electromagnetic fields: AC-field induced ponderomotoric forces," *British Journal of Cancer*, vol. 45, suppl. V, 1982, pp. 220–224; and "Research on biological effects of nonionizing radiations: contributions to biological properties, field interactions and dosimetry," *Bioelectromagnetics*, vol. 7, 1986, pp. 113–128.

[69] Steneck et al., "The origins of U.S. safety standards . . . " and Nicholas H. Steneck, *The Microwave Debate* (Cambridge MA: MIT Press, 1984), pp. 55–60.

[70] Interview 1992, pp. 85–86.

[71] Steneck, *The Microwave Debate*, pp. 61–62.

[72] Telephone interview of Schwan, 30 October 1992.

[73] Steneck, *The Microwave Debate*, pp. 189–195.

[74] Paul Brodeur, "A reporter at large: microwaves-I and -II," *New Yorker*, 13 and 27 December 1976, and Paul Brodeur, *The Zapping of America* (New York: Norton, 1977).

[75] See U.S. Senate, Committee on Commerce, *Radiation Control for Health and Safety Act of 1967, Hearings on S. 2067, S. 3211, and H.R. 10790*, 90th Congress, 1st, 2nd Sessions, 1967–1968, pp. 699–718; and *Congressional Record*, 93rd Congress, 1st Session, Hearings for the Committee on Science and Astronautics, 1973.

[76] See the quotations of Schwan on page 125 of Steneck's *The Microwave*

Debate.

[77] Schwan, "Entwicklung. . . ."

[78] Telephone interview of Schwan, 30 October 1992.

[79] *EPRI Journal*, March 1992, pp. 5–13.

[80] Interview 1992, pp. 98–102.

[81] Interview 1992, p. 205.

[82] "Biomedical engineering and basic research: historical perspectives," *Medical Progress through Technology*, vol. 7, 1980, pp. 63–67; "Physical properties of biological matter: some history, principles, and applications," *Bioelectromagnetics*, vol. 3, 1982, pp. 3–8; "Ultrasound and electromagnetic radiation in hyperthermia—a historical perspective," *British Journal of Cancer*, vol. 45, suppl. V, 1982, pp. 84–92; "Historical review, state of the art, open problems," in A. Chiabrera, C. Nicolini, and H. P. Schwan, eds., *Interactions between Electromagnetic Fields and Cells* (New York: Plenum Press), pp. 1–18; "Early history of bioelectromagnetics."

[83] "Microwave bioeffects research: historical perspectives on productive approaches," *Journal of Microwave Power*, vol. 14, 1979, pp. 1–5; "History of the genesis and development of the study of effects of low energy electromagnetic fields," in *Advances in Biological Effects and Dosimetry of Low Energy Electromagnetic Fields* (New York: Plenum Press, 1983); "Microwaves and thermoregulation: historical introduction," in *Microwaves and Thermoregulation* (New York: Academic Press, 1983); "Science and standards. RF-hazards and standards: an historical perspective," *Journal of Microwave Power*, vol. 19, 1984, pp. 225–231.

[84] "The development of biomedical engineering: historical comments," *IEEE Frontiers of Engineering and Computing in Health Care* (Proceedings of the sixth annual conference of the IEEE EMB Society, 1984), pp. 851–857; "The development of biomedical engineering: historical comments and observations," *IEEE Transactions on Biomedical Engineering*, vol. BME-31, 1984, pp. 730–736; "Organizational development of biomedical engineering" (cited above); "Biomedical engineering at the University of Pennsylvania" (cited above); "Entwicklung der Biophysik und das Frankfurter Institut für Biophysik" (cited above).

CHAPTER 3

From Radar Bombing Systems to the Maser
Charles Townes
as Electrical Engineer

CHARLES TOWNES

Figure 1. Charles Hard Townes, Nobel Laureate, National Medal of Science winner, recipient of the Morris Liebmann Award of the Institute of Radio Engineers and the Medal of Honor of the Institute of Electrical and Electronics Engineers, and currently professor emeritus at the University of California, Berkeley.

The summer of 1939 found most Americans in an optimistic mood. Signs of economic growth suggested that the Great Depression was finally ending. World's fairs in New York and San Francisco—featuring such marvels as electronically synthesized speech, artificial lightning, and a cigarette-smoking robot—showed millions how new technology might improve transportation, communications, housing, and leisure. Fluorescent lighting, nylon stockings, magnetic tape recording, and television were products of the future displayed that summer. Pan American Airways began regular transatlantic passenger service on June 28 of that year and the first FM radio station (Edwin Howard Armstrong's W2XMN of Alpine, New Jersey) began regular broadcasting on July 18. But events in Europe, long troubling to many, rather suddenly had a sobering effect on everyone: On August 24 Americans learned of the German-Soviet nonaggression pact (seeming to untie Hitler's hands), on September 1 Germany invaded Poland, on September 3 England and France declared war on Germany, and on that same day thirty Americans lost their lives when a

German submarine sank the British passenger ship *Athenia*. On September 5 the US government announced its neutrality, and on September 8 President Roosevelt proclaimed a limited national emergency.

Starting Work at Bell Telephone Laboratories

It was in this somber atmosphere, at the beginning of September 1939, that a group of new employees were welcomed at the Bell Telephone Laboratories at 463 West Street in New York City. A dozen or so newly minted Ph.D.s were given an introduction to the wide-ranging work of the largest and most prestigious industrial research establishment in the world. The new employees were also of the highest quality: Two of them, James B. Fisk and William O. Baker, rose within the organization to serve long terms as President of Bell Laboratories, and a third, Charles Townes, became famous as inventor of the maser and co-inventor of the laser.[1]

Earlier that summer Townes had completed his physics dissertation at the California Institute of Technology and used the first-class train fare sent him by his new employer to travel from California to New York to detour through Guadalajara, Mexico City, and Acapulco by traveling third class. Townes was to continue in an exploratory mode at Bell Labs, as Mervin Kelly, Director of Research, arranged for him to work for a few months in each of several departments, with the idea that this would help Bell Labs and Townes decide what work he was most suited to.

Townes was first assigned to the group of Frederick B. Llewellyn, an engineer who three years before had received the Morris Liebmann Award of the Institute of Radio Engineers for his work on high-frequency electronics and constant frequency oscillators and who, in 1946, became President of the Institute. Llewellyn was then trying to design good generators of microwaves, ultra-high-frequency radio waves with wavelengths below 100 centimeters. At that time, triode vacuum tubes (especially of the Barkhausen-Kurz type), klystrons, and magnetrons were the principal means of generating microwaves, and despite a great deal of work at laboratories in the United States, Europe, and Japan, it was proving extremely difficult to build an oscillator of short wavelength and high power.[2]

Townes undertook to explain theoretically why certain tubes, which were being tested by Llewellyn's group, performed as they did. For this work he had the advantage of an unusually thorough training in electromagnetic theory at Cal Tech from W. R. Smythe, the author of a standard text on the subject. Smythe gave an extremely demanding electromagnetic theory course to all physics graduate students and was proud of the fact that this course "winnowed out everybody who couldn't quite take it at Cal Tech."[3] Townes not only excelled in the course, but ended up helping

Smythe by checking all the derivations and working all the problems in a new book Smythe was writing.[4] Townes, who thus gained a facility with electromagnetic calculations that few engineers had, has commented, "So I knew that field. I think once you know one basic field very thoroughly, that's an enormous help in almost anything you do. . . . you have a very powerful tool. And, yes, that has been very important to me. It was certainly one of the things that gave me a start at Bell Labs."[5]

[Electromagnetic theory] touches on so many fields. I think learning that field as thoroughly as I did has really been quite important to my career.

There's a more general aspect that I would comment on: I think it's very important to learn a subject well enough, to think about it enough, to think about it from all angles, so that you feel at home with it. You feel it's a friend. Something you know intimately. If somebody asks you about it, you can express it back to them in your own way or anybody else's way. You can look at it from any angle and still understand it thoroughly. This gives one a distinctive idea of what happens, and what's going to happen if you're given a certain set of events. You just see it. You visualize what's going on. It's instinctive.

Now I always work out things with equations, too, to be sure it's all correct, to be sure I haven't missed anything. I usually think things through intuitively first, work it out with equations, and then think through the equations: Now, what do these equations really mean? Is that really what's happening? Do these equations really describe the right physical situation? I think the intimacy you have with a given field is very important for quick thinking and also for exploring a problem thoroughly.

Teaching is another way of learning thoroughly, and I have profited from that. I think if you teach a field, then you have to look at it in any ways your students want to look at it. You have to look at all aspects of it. If you just learn it in other ways, you may well skip over certain parts you don't think are too interesting. If you've got to teach it, you have to look at all of those. And I have gotten some very good ideas just having to learn certain parts of a subject which I've otherwise not worried about, and I'll suddenly see something that's there that I'd missed.[6]

In late October 1939 Townes had an idea for a new way to generate high-frequency waves, using nonuniformity of an oscillating electric field to transfer energy from electrons to the field.[7] He thought this invention had promise and had its notebook description witnessed. He later said, "I did not try it out. I'd moved on to another department by then. But Jim Fisk, who was in the tube department, was asked to try it out. Jim tried it out and said it didn't work. So that was the end of that. . . . I'm sure it would have worked in some sense; but if it didn't work . . . more or less right away, then that probably meant that it wasn't all that useful."[8]

Though not in the business of making money from patents, Bell Labs worked hard to protect itself against infringement suits by taking out patents on its inventions. While at the Lab, Townes followed company practice, signing and dating notebook entries that might someday provide evidence in a patent case and informing the Lab's patent office of promising ideas. For work done at the Lab, Townes received a total of ten patents.[9] "Since then I've patented practically nothing."[10] (He did patent the maser, but, as explained below, turned the patent over to Columbia University; he and Arthur Schawlow received a basic patent on the laser, but Bell Laboratories held this patent.)[11]

In early December Townes reported to J. R. Wilson, Director of Vacuum Tube Development, and began work with the engineer G. H. Rockwood to understand cathode sputtering, the gradual disintegration of the cathode (through positive-ion bombardment and through evaporation of atoms) that in many cases determines the lifetime of a vacuum tube. Townes analyzed some data taken by Rockwood and helped collect other data.[12] The work required theoretical insight (for example, into the work-function of a metal surface), engineering expertise (as in knowing that "torque in a quartz fiber = $\frac{\pi Z r^4 \theta}{2L}$"), calculational ability (framing problems in a way that permits calculation and knowing techniques of approximation), and experimental ingenuity (as building "a modification of the Knudsen gauge"). He later commented, "Well, I would say I enjoyed this kind of work. It's the type of engineering which is so close to physics. It's not development of a big system . . . which I later did. Rather it involves trying to understand things."[13] As discussed below, he returned to similar work several months later.

At the beginning of March 1940 Townes moved to his next assignment, the magnetics group headed by the physicist-engineer R. M. Bozorth. Bozorth had come to Bell Labs in 1923, and his work since then had helped make it the world leader in the understanding and exploitation of magnetic materials. According to a 1969 editorial in *IEEE Transactions on Magnetics*, Bozorth did "more to promote the interaction between research and engineering in magnetism than any other person of our time."[14] Townes worked closely with Bozorth and learned much in this period.[15]

In mid-April Townes was asked to report to Dean Wooldridge, who was studying electron physics, such as secondary emission of electrons from surfaces. (Wooldridge too had earned his Ph.D. under Smythe at Cal Tech and had come to Bell Labs in 1936; he went on to become head of TRW, Thomson-Ramo-Wooldridge.) The exploratory period of Townes's employment at Bell Labs here came to an end as he undertook a long-term project. The objective, set by Harvey Fletcher, Director of Physical Research, was to design a gas discharge tube that would operate at 24 volts. The gas discharge tubes then available required considerably higher voltage, which made it impractical to power them by batteries, and there were places in the Bell System where such tubes could be used as switches if they could run on batteries.

So Townes set to work studying the fundamental mechanisms of gas discharges in order to learn how to achieve breakdown at lower voltage.[16]

The work was similar to Townes's earlier work with Rockwood. Indeed, Townes continued the investigation he started there of cathode sputtering; this resulted in a Bell Labs Technical Memorandum in January 1941 and, several years later, in a paper in *Physical Review*.[17]

The work on a 24-volt gas discharge tube was in part theoretical—as in using quantum mechanics to calculate the probability of ion capture of an electron—and in part experimental, involving the testing of new tube designs. The work continued until March 1941, when it was abruptly terminated. Though his notebook testifies to diligent effort and presents both theoretical and experimental results, Townes says, "... I never made very much progress in this discharge tube business."[18]

Figure 2.   Charles Townes in 1941.

Because of the great variety and sophisticated nature of the research going on at Bell Labs, Townes found it a very stimulating environment. He learned, for example, of two different efforts to carry out calculations automatically. In 1937, George Stibitz had begun designing an automatic calculator that would use existing telephone components: relays, sequence switches, and teletypewriter equipment.[19] Townes recalls: "I remember very well his rolling around a relay rack . . . to show people how you could compute with relays. He said, 'Look, you've got to do it digitally. Otherwise you'll never get any precision. You have to do it digitally.' Well, he was doing it with relays." Chuckling, Townes continues: "A rack full of mechanical things. . . . People were not highly impressed with that."[20] Stibitz's work continued, however, and led, before the end of the war, to several digital computers that were used in the design and testing of analog gun directors.[21]

To most people at the time, the work that had recently begun at Bell Labs on electrical analog computing was more impressive. Mechanical analog calculators, such as slide rules, planimeters, and Lord Kelvin's tide predictor, had long been in use, and in the early 1930s Vannevar Bush at MIT had added electrical control to mechanical calculation in his "differential analyzer."

Beginning in 1940, the Bell Labs engineer D. B. Parkinson and his supervisor C. A. Lovell—who were trying to design an automatic gun director—developed what they called "electrical mathematics." By representing variables by voltages, they showed how electrical circuits could add, subtract, multiply, divide, differentiate, integrate, and even make use of tabulated data.[22] A network of resistances could be used to sum voltages, and an amplifier could be used to multiply a voltage by a constant. Specially designed potentiometers became a basic component of their system, as Townes explains: "Lovell had the idea of shaping a potentiometer so that for a given angle you could get a more arbitrary function. You wound the potentiometer on a card of varying height so that the resistance varied, not linearly but with some other kind of functional form."[23] As we will see, Townes came to be very much involved in Lovell's work.

Now one other thing I would have to say about Bell Labs: I think it had a very good atmosphere in not differentiating between physics and engineering in any strong clear-cut way. They had a physics department, and they had various kinds of engineering departments. They had a chemistry department. But there was a lot of interaction. Engineers would work in the physics department a bit, and vice versa. There was a good deal transferred back and forth, and a good deal of interplay. There was not a sharp distinction between the two. . . . [Bell Labs] made a practice of hiring physicists and then transferring them into engineering. During that period engineering schools did not teach a lot of fundamental physics. Cal Tech was one of the few, and Cal Tech did it simply because it didn't have much of an engineering faculty. [Laughs] . . . I was a student there, and the engineering students would take a lot of fundamental physics because they had a small engineering faculty. . . .[24]

It was also at about this time that Mervin Kelly, Director of Research, started a seminar for a small group of employees to keep abreast of the latest developments in physics, engineering, and chemistry. Townes felt privileged to be selected; others were William Shockley and Walter Brattain (later famous as co-inventors of the transistor), James Fisk, Dean Wooldridge, and Stan Morgan (who later headed chemical research at Bell Labs). ". . . [we] were to meet together once a week and basically have an afternoon off to talk about some aspect of physics or related matters. We could do anything we wanted to. We could read scientific papers, we could invite somebody to come and talk, we could read through important new books or something like this. Whatever we wanted. And the Laboratory provided tea and cookies, which was again quite unheard of at that time."[25]

Townes found his life outside the laboratory stimulating too. He enjoyed New York theater, museums, and restaurants, and, eager to get to know the city, decided to move every three months or so. He lived first in Greenwich Village, then uptown near Columbia University, then in midtown Manhattan. "I would simply put all my stuff in a trunk, get a taxi, move to a new place, and get acquainted with the neighborhood."[26] He took voice and music theory lessons at Juilliard. On a ski trip he met Frances Brown, who was working as activities director at the International House of Columbia University, and the two were married 4 May 1941. (In 1991 they celebrated their golden wedding anniversary along with their four daughters and six grandchildren.) (See Figure 3.)

Figure 3. A 1956 photograph of Charles and Frances Townes with their daughters (from left to right) Linda, Carla, Holly (seated on the floor), and Ellen.

Radar Bombing Systems

In April and May 1940 Germany's blitzkrieg overwhelmed opposition armies on the Continent, and Britain, mortally threatened by German bombers and submarines, stood alone in its fight against Hitler. It appeared more and more likely that the United States would be drawn into the conflict. In May President Roosevelt named a National Defense Advisory Commission to coordinate civilian and military defense, and in June he established the National Defense Research Committee, which was charged with mobilizing the nation's scientists and engineers—whether in industry, academia, or the military—for national defense. In September the United States sent 50 destroyers to England in exchange for air and naval bases in Newfoundland and the West Indies, and the first peacetime draft in US history began.

Bell Labs had not undertaken military R & D until 1937 when the Navy asked the laboratory to investigate the use of radar for an automatic gun director. Over the next several years the number and range of military projects increased, including work on radar systems, specialized communications systems, sonar, the proximity fuse, the acoustic torpedo, and magnetic detection of submarines. In 1940, military projects accounted for 2.5 percent of the Bell Labs' budget; this grew to almost 85 percent in the following two years.[27] Nearly all research work not related to the war effort had to be set aside.

Townes remembers being called into a meeting with Mervin Kelly at the end of February 1941. There he was told that on the following Monday he would start work on a radar bombing system, as part of a group headed by Dean Wooldridge and including Sidney Darlington and one or two others. Townes had received no warning that a change of assignment was to be made: "... that was pretty sudden and unexpected, and I wasn't accustomed to being treated that way."[28] Nevertheless, he did not resent the action, believing that US involvement in the war was imminent and that to contribute to national defense was everyone's duty.[29]

So on Friday, 28 February Townes closed out Notebook 17015 with a page headed "This work discontinued for national defense job," which listed all of the apparatus that was put into storage.[30] The following Monday he made the first entry in Notebook 17870, which was a diagram and analysis of a servomechanism. The assignment of Wooldridge's group was to design, build, and test a bombing system that would make use of radar and the analog computing techniques being developed by Lovell.

Bombing played a large part in the Allied victory in World War II, and its effectiveness was highly dependent upon the techniques used to drop the bombs accurately. The Norden optical bombsight (itself a mechanical analog computer) gave satisfactory results under certain conditions: daylight, clear weather, and little danger from antiaircraft fire or enemy fighter-aircraft (in order to permit a straight run at constant speed to the release point). To achieve accuracy under other conditions, military planners turned to electronic bombing systems, consisting of electrical analog computers and radar devices.

Most of the bombing radars worked as follows: the bombadier identified the image of the target on the cathode-ray screen of the radar set, moved a set of cross-hairs to cover the image, and then kept the cross-hairs on the image as the plane approached the target; an analog computer took as input such things as altitude of the airplane, ground speed, winds, and ballistics of the bomb, and gave output in the form of a needle showing the pilot which way to turn and an indication to the bombadier when to release the bomb.

Such a bombing system was a combination of two new technical devices: an airborne radar set and an electrical analog computer. Bell Labs, as we have already seen, was the leader in electrical analog computing, and it had already established its expertise in radar. (In the United States the design of almost all World War II radar systems was done either at Bell Labs or at the MIT Radiation Laboratory.) The previous September a group at the Bell Laboratories in Whippany, New Jersey, had begun work on a bombing system that combined an S-band radar (that is, one producing 10-centimeter microwaves) with a version of the analog computer developed for directing antiaircraft fire.[31] Wooldridge's group was to design and build the computer for a new system, while the radar set would be designed and built by a group in Whippany.

Townes took part in the conceptual design of the system, such as what variables were to be used as input and how they were to be measured. The wind speed, for example, was read from a more or less standard pitot tube and corrected by taking into account air density and temperature. Townes took part also in the hardware design. The basic calculating device was the custom wound potentiometer, and in the design of each of these one had to consider accuracy and the amount of current drawn. He had to be concerned also with the design of phase-shifting devices, of circuits for controlling motor speed, and of servomechanisms (as for automatically keeping the radar antenna aimed at the target as the plane maneuvered). His training in mechanics was invaluable in calculating such things as torque required to move a worm gear, efficiency of transmission of torque, and the condition's needed to avoid the binding of worm gears. Townes took part also in the testing of components and of the system as a whole.[32] There were many subjects, such as amplifiers and servomechanisms, that he learned thoroughly for the first time in doing this work.

The group worked rapidly designing, building, and testing a prototype system. (The circuit shown in Figure 4 was designed in the first six weeks of the project.) The following winter, shortly after the Japanese attack on Pearl Harbor had compelled US entry into the war, a unit had been built and was ready for in-flight testing. In February and March, Wooldridge and Townes attended the unit on some fifty trial bombing runs out of Boca Raton, Florida.[33] Though the system was designed to require little human intervention, Wooldridge and Townes needed to be present to adjust and repair the system and to observe its operation.

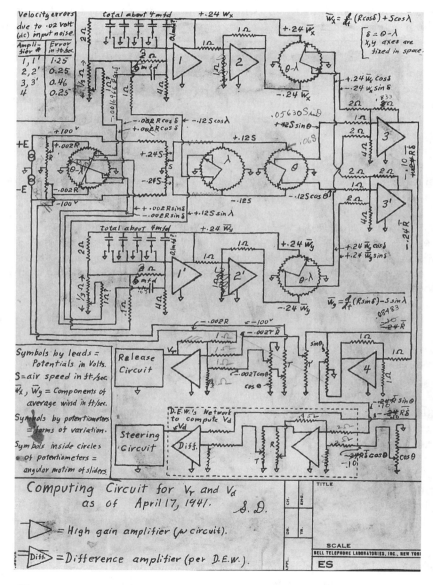

Figure 4. A diagram, drawn by Sidney Darlington, from Townes's notebook showing the computing circuit for two of the variables used in the bombing system.

We did our first bombing exercise, bombing a ship that was anchored offshore. One of the curious things is that the Norden bombsight was so highly thought-of at that time—and that made it so secret—that even though we were supposed to be designing a bombing system they would not tell us anything about the Norden bombsight. We couldn't see it, we couldn't know what it did. We asked our Air Force representatives, "What kind of precision do you get? What kind of precision do you need?" They would just say, "Just do the best you can." People talk about the Norden bombsight dropping bombs in pickle barrels, and they had great stories about it. But the actual accuracy was not all that good. . . .

The very first run we had, we had a colonel who was a very nice person. . . . [He] flew us, and the first bomb we dropped we had a run at an altitude of five-thousand feet. Five-thousand feet was a reasonably high altitude at that point. We dropped the bomb, and I quickly dashed up to the cabin to see what happened. It missed by about a hundred feet. And this colonel, who wouldn't tell us anything about the Norden bombsight— yes, sure, he'd used it. He couldn't say a thing about it. So then he said, "That's a damned good shot, if you ask me." That boosted our morale a great deal, and gave us our first real information on the accuracy then.[34]

TOWNES: *. . . [the Army Air Corps] would send people from time to time to fly with us and see how it was coming out. That was not so unusual. One time we had a load of fairly high-level people— some Bell Labs people and some people from Washington—fly in our plane. The radar antenna housing got stuck. That is, you let the antenna down out of the body of the plane so the antenna could see out, and then you'd raise it back up where the antenna couldn't see, in order to land. Well, it was let down, and we couldn't get it up. Everybody thought that all these bigwigs were going to have to jump. [Laughs] Oh, dear, what had we done now. All these older men, distinguished people, we were going to make them jump out of there.*

NEBEKER: *Couldn't you land the plane anyway?*

TOWNES: *No, they said it was too dangerous. Too dangerous. So I managed to climb down in there with some wrenches and screwdrivers and get it fixed. It was generally motor-driven, you see, and it had gotten jammed, and they couldn't do a thing with it. It was a heck of a job. Did it by hand and got it back up again.[35]*

Over the next four years Townes divided his time between the New York area (at about this time he moved from the Bell Labs offices on West Street to the new campus in Murray Hill, New Jersey) and Florida. The testing of a bombing system usually took a month or two. Most flights were from a base at Boca Raton; others were from a base near Pensacola. As the quotation above suggests, this was a dangerous business. Though Townes was not involved in any serious accidents, three of the four Air Corps pilots he flew with died in noncombat airplane accidents before the end of the war.

In the spring of 1942 Wooldridge and Townes began work on a new system (designated the D-150550 bombsight radar), whose radar used 3-centimeter microwaves. During the war there was a continued move toward shorter wavelengths: from 23 centimeters (L band) to 10 centimeters (S band) to 3 centimeters (X band) to 1.25 centimeters (K band). Shorter wavelengths allowed improved resolution and reduced size of components (notably the antennas). Wooldridge's group had first worked with S-band radar; now they were responsible for the overall system design ("marrying the radar and the computer"[36]), as well as the computer design, of a bombing system using X-band radar. (The radar itself was designed by another Bell Labs group.)

While ships and coastlines were readily discernible with S-band radar, bombing over land usually required higher resolutions for target identification. The new system promised improvement in this respect by employing X-band radar. The other main advantage of the new system was that it would not require level flight, so the pilot could change elevation as he approached the target.[37] In addition, Wooldridge and Townes designed a version of this system for use in torpedo bombing.[38]

Other refinements were made to this and later bombing systems: allowing optical as well as radar guiding (so that a break in the cloud cover could be used to advantage), making input of certain variables automatic, taking into account new variables, and allowing for "offset bombing" (where the target is different from the ground site used for tracking, since not all targets show up well on radar).

For certain functions, the accuracy achievable with the custom-designed potentiometers was not sufficient. Townes worked, for example, on a circuit to provide a better approximation of $x \sin (y + z)$.[39] He also devised a novel way to get a better approximation of the cosine function: He modified a potentiometer output by using vacuum tubes to switch the circuit, when it was in a particular voltage range, to an appropriate subcircuit.[40] Colleagues were surprised when Townes showed that this fairly complicated way of calculating the cosine actually worked.

All of the systems Townes worked on were quite complicated, incorporating many tubes, potentiometers, motors, and mechanical relays. The tubes, because they were being used in calculations, had to have a linear and accurate response, and on at least one occasion the group had a special tube manufactured to its specifications. With such components, there were frequent problems: "... by the end of the war I felt we had as many [tubes and relays] as we could afford. Otherwise, the troubles with them would be too frequent."[41]

In this work Townes learned systems engineering. His notebooks include long lists of aspects of a system to be checked or problems to be solved, such as "Tilt motor did not quite have sufficient power to run drive,"

". . . amp[lifier] #10 is off balance," "Antenna azimuth drive has been jittery and oscillating on occasion—usually just during warm up."[42] (See Figure 5.) Some pages of the notebooks consist entirely of lists of problems or of adjustments and corrections or of tests to be done. Looking at one of these pages in 1992, Townes remarked, "Well, I'm still doing that kind of thing with our telescopes that I'm currently working with."[43] (After a period in which he had been working only on relatively small laboratory instruments, Townes submitted a proposal to the National Science Foundation for a large and complex telescope system. In Townes's words, "One of the reviews came back saying, 'Well, can Townes build anything like that?' [Laughs] I guess he didn't know about my war work. There was a good deal of similarity."[44])

SECRET.   SEP   1 1943

Matching Resistance of Cards 646642 and 646643 - Case 23665-1

Mib

September 1, 1943-1115-CHT-GM

GROUP 4
Downgraded at 3-year intervals;
declassified after 12 years

MR. L. N. HAMPTON:

        Subsequent to your letter to Mr. D. E. Wooldridge
of August 14th questioning requirements on the resistance
ratio of α_1 to α_2 cards of the D-164558 potentiometer,
these requirements have been reviewed and recalculated.
This situation was discussed with Mr. W. W. Werring when
the circuits for α_1 and α_2 were designed so that we were
aware that a close tolerance on the resistance ratio of
these two cards is a difficult manufacturing problem.
However, the circuit simplification afforded by this design
seemed to justify its difficulty.

        Since the original specification of these cards,
it has become possible to broaden the tolerance on the
resistance ratio of any pair from ± 1% to ± 3% of nominal
value and still obtain the desired accuracy. It is hoped
that this will facilitate manufacture and that further
increase in tolerance with consequent decrease in accuracy
will not be necessary.

                        C. H. TOWNES

Copies to
Messrs. E. T. Mottram
        W. W. Werring
        D. E. Wooldridge

W.J. Fritz advised o/r by W.W.W.
JH .

Figure 5.   One type of problem Townes had to deal with is illustrated by this letter from Townes to L. N. Hampton, who was in charge of manufacturing the potentiometers used in the system Wooldridge's group was building.

Throughout the war, work proceeded at an intense level at Bell Labs. According to a history of the laboratory, "Early in 1942, the normal 48-hour work week was increased by nearly 40 percent, and the actual hours worked often went far beyond...."[45] Though corroborating this, Townes reports that he often found himself with short intervals of leisure because there were frequent delays in testing the bombing systems, as when the weather was unsuitable for flying or the plane had mechanical problems. In many of these intervals, he worked at solving a scientific puzzle he had set himself.

In 1930 Karl Jansky of Bell Labs was asked to investigate "atmospherics," disturbances to radio transmission believed to be caused by electrical phenomena in the atmosphere. In a 1933 paper Jansky argued that some of the electrical disturbances were of extraterrestrial origin.[46] Already as an undergraduate Townes had read of Jansky's discoveries and wondered how radio waves might be generated in space. He finally succeeded in finding a plausible mechanism for the generation of these waves and soon thereafter published his theory of the so-called free–free electron transitions, which was the first to explain correctly how radio waves were emitted by galactic ionized gas.[47] This work may have owed something to his efforts, in his first assignment at Bell Labs, to devise a way to generate microwaves.

None of the systems Townes worked on were used during the war, and this bothered Townes. (Quite a few other groups at Bell Labs developed bombing radar systems, and a number of these went into production before war's end.[48]) He would rather have seen one of the systems rushed into production and put to use, whereas it seemed that the completion of one design led only to requests for still more complicated systems operating at shorter wavelength with improved performance and new capabilities.[49] He considered leaving Bell Labs. "I was thinking ... of going over and joining [General Joseph W.] Stilwell in China as some kind of technical assistant there.... My bosses then got wind of this, and they worked on me very hard to stay. I didn't quite see the right kind of an opening there, so I said, 'Well, okay. I'll just keep on doing this.' "[50]

Townes's next assignment was, not surprisingly, a still more advanced bombing system, one incorporating the new K-band radar. On the one hand, this assignment proved even more frustrating: Townes could see that it would not work effectively, but was unable to convince his supervisors of that. On the other hand, this assignment led directly to his path-breaking work in molecular spectroscopy and from there to the invention of the maser and laser.

In September 1944—when Allied armies, just three months after D-Day, reached German soil and it appeared the war in Europe might be over by Christmas—Wooldridge and Townes were asked to begin work immediately on a system (the AN/APQ-34 high-altitude bombing radar) that had already been designed at the MIT Radiation Laboratory. According to Townes, "That was a little bit of a blow ... because they had suddenly tried to jump ahead of us and do something still more advanced than we were doing, and had sold it to the military, and the military had said, 'Well, of course, if we're going to have it manufactured, that ought to be AT&T....' "[51] Bell Labs was to redesign the system for manufacture and build a prototype.

Substantial redesigning was required, and as part of the process Townes did a theoretical analysis of the total error of the system. He obtained estimates of the error for each variable of each component (which took five pages to tabulate), calculated the error in output of each component (expressed as milliradians error at an elevation of 2000 feet), and (assuming each error distribution to be Gaussian and independent) calculated total error as the root mean square of component errors.[52] Today this is standard practice, but at the time Townes felt he was being quite innovative. (In test flights the actual errors in range and deviation were each typically one or two hundred feet; this compared favorably with what the Norden bombsight could achieve in daylight with good visibility, conditions not of course required for the radar systems.)

While it was easy to allow bombardiers the choice between visual and radar bombing—by providing both a Norden bombsight and a radar system—it was difficult to design a single system that could take advantage of both visual and radar input. This was one of the objectives of the new system, and Townes considered various ways of integrating visual sighting and radar sighting.[53]

Townes, however, soon came to have doubts about whether the radar system would work. The new K-band radar seemed to offer improved resolution, but it was untried. A memorandum written by the physicist John Van Vleck alerted Townes to the possibility that water vapor might absorb strongly in exactly that frequency range.[54] After further calculations, he decided that the absorption would be so strong that the radar would be unusable in applications, such as bombing radars, requiring much range. Townes explained his doubts to others at Bell Labs, to I. I. Rabi at the MIT Radiation Laboratory, and to the Pentagon, but without result. "I tried to persuade people that it wasn't going to work, but I was too young and the decision had been made. Well, they went ahead and put it in the field, and it had no range because of water-vapor absorption. So all the equipment was junked."[55]

One thing I remember is that when the first atomic bomb was dropped [6 August 1945]. . . . I was working [late] in a little hut in Whippany checking out some radar. I've forgotten what. But I remember this little hut, and I had a radio, and I heard that this bomb had been dropped. And it was an unknown kind of bomb that did a very powerful job. Well, I knew precisely what it was. A number of friends I had who had been working on the system were indiscreet enough to keep me posted on what was happening. I remember very well. And I said, "Well, I don't have to keep working today." So I shut down and went home.[56]

Microwave Spectroscopy, the Maser, and the Laser

When the war ended, Townes was eager to return to more fundamental research. Like many physicists, he believed that his wartime engineering experience would be of great benefit in scientific research. He considered investigating the radio waves coming from space, but was advised by knowledgeable people that one could never get much information from those waves.[57] Townes might have pursued that possibility nevertheless—and in the next few years a new field of radio astronomy did emerge, which was pioneered mainly by people who had worked on radar during the war[58]—had he not seen another promising avenue of research.

His work on the K-band radar system got Townes thinking about the interaction of microwaves and molecules. With the generators of microwaves and the sensitive detectors that were developed for radar, one could, Townes thought, do a powerful new type of spectroscopy, one that would determine molecular properties with much greater accuracy than any method available at the time. He had calculated that as the density of a gas decreased, the width of the absorption line would decrease but—contrary to what spectroscopists expected—the intensity would not. Townes recognized that the ability to record the sharp absorption lines would give one precise information about molecular structure, such as bond distances and bond angles.

Townes saw also that microwave spectroscopy, as he called it, might be of value to a communications company such as AT&T. In a memo he wrote to his supervisors arguing for such a research program, he stressed the possible applications while admitting that, "So little work has been done in microwave spectroscopy . . . that the course of its future development and application is difficult to predict."[59] His general argument ran as follows: "Microwave radio has now been extended to such short wavelengths that it overlaps a region rich in molecular resonances, where quantum mechanical theory and spectroscopic techniques can provide aids to radio engineering. Resonant molecules may furnish a number of the circuit elements of future systems using electromagnetic waves shorter than 1 cm." In the memo Townes listed possible applications of the knowledge of molecular resonances, including chemical analysis, detection of radio waves (using molecules as detectors of the frequencies corresponding to their resonances), establishment of frequency standards, and generation of high-frequency waves. In a laboratory notebook he was more specific about what circuit elements might be based upon molecular absorption and emission: frequency discriminators, band absorption filters, band pass filters (by re-emission), time delays (by absorption and re-emission), and oscillators.[60]

The memo was well received, and as soon as Townes found someone to take his place on the radar project—this did not happen until early January 1946—he was allowed to begin work on microwave spectroscopy (and was provided with two technical assistants). He first studied the absorption, at various pressures, of water vapor and ammonia. This rapidly led to other fruitful investigations, and over the next decade he published an average of ten articles per year. For this work and for a classic text, *Microwave Spectroscopy*, he is recognized as one of the founders of the field. *Microwave Spectroscopy*, coauthored with Arthur Schawlow, was first published in 1955 and is still in print.[61]

Townes has pointed out the dependence of this new field on electronics generally: "Electronic techniques are characteristic of microwave spectroscopy, being involved in the production, detection, and amplification of microwaves. In some cases very sensitive electronic circuits are needed for proper detection and amplification, since the fractional power decrease may be quite small—as small as one part in 10^8 in an absorption path of 1 meter."[62]

The K-band radar system was important to this development in two ways. Though Townes had some earlier experience with visual spectroscopy, it was the K-band system that directed his attention to molecular spectroscopy. Secondly, the K-band equipment was just what was needed for the new field, and because the military was no longer interested, the equipment was sold as surplus. Townes recalls, "there was suddenly abundantly available lots of Klystrons and wave guides and equipment and detectors at this wavelength.... In fact, I remember buying a few Klystrons down on the sidewalk in lower New York City...."[63]

Townes was delighted to be doing scientific research again, but he had not abandoned engineering. Besides doing the engineering required for the spectroscopic apparatus, he was still alert to technological possibilities: In August 1947—in the midst of this flurry of work on molecular spectroscopy—he took time to pursue an idea he had for a new type of television picture tube (based on the phenomenon of bombardment-induced conductivity) and even obtained a patent on the scheme.[64]

Though Townes was happy at Bell Labs and was not seeking another job, when I. I. Rabi offered him an associate professorship at Columbia University, he accepted, beginning work there in January 1948. He preferred a university setting: ". . . I like teaching, I like interaction with the students. I also like the breadth of the university, having people in the humanities and other fields that you can interact with."[65] He was also disappointed that Bell Labs had declined his request that an additional researcher in molecular spectroscopy be hired.[66]

At Columbia Townes continued using microwave spectroscopy to determine properties of molecules, atoms, and nuclei. But he began to search for ways to expand the field. "After about 1948 . . . the backlog of wartime techniques was beginning to run out. I felt the need for still shorter wavelengths—generated coherently by oscillators. So from time to time I tried to think of ways one might generate shorter and shorter waves, by harmonic generation and new types of tubes and so forth. This got me back into applied work but for the purpose of producing tools for basic research."[67] The method that seemed to him most promising, and to which he gave most of his attention, used Cerenkov radiation to generate the short wavelengths.[68] But in 1951 he had an idea for a new type of oscillator that would succeed beyond everyone's expectations.

In 1950 the Navy had asked Townes to head a committee to examine the possibilities of new techniques for generating millimeter waves, and in April 1951 he was in Washington D.C. to chair a meeting of this committee. Waking up early that Sunday morning, he went for a walk, and while sitting on a bench in Franklin Park he had the idea for a new way to generate millimeter waves. If a gas could be produced in which most of the molecules were in an excited state, then electromagnetic waves of just the right wavelength would be amplified by stimulating the molecules to emit energy at that wavelength.

Townes, working with graduate student James P. Gordon and post-doctoral student Herbert J. Zeiger, finally succeeded in getting such a device to work in April 1954. Called the maser (**m**icrowave **a**mplification by **s**timulated **e**mission of **r**adiation), it was soon being used as an amplifier (about one hundred times more sensitive than preexisting techniques) and as an oscillator (of such constancy that a maser clock could be accurate to 1 second in 300,000 years).

After remarking that the physics literature already contained mention of the possibility of obtaining amplification by stimulated emission, Townes pointed to two essential contributions of his group: recognizing the coherence of the resulting radiation (more readily conceived with the engineer's view of radiation as consisting of waves than with the physicist's view of radiation as photons) and the adding of an external resonator for feedback.[69] Townes and other physicists had considered the possibility of getting stimulated emission from an inverted population, but the weakness of that effect made any practical application seem highly unlikely; feedback in a resonant cavity is what made masers practical.

In his years at Columbia, Townes's engineering skills were called upon constantly: "I personally [assembled] the apparatus mostly. . . . I was the one who knew all the parts of the apparatus and how it would go together, and I fixed the leaks and did everything with the electronics, with the students helping me. That's rather different from the way I have to work now. Mostly they are the ones who know the apparatus. . . ."[70]

Townes, along with many others, sought ways to extend the frequency range of the maser. He believed that what was needed was a high-Q resonator having only a few modes of oscillation (Q is a ratio of energy stored to energy dissipated), and believed that this could be done more easily by jumping over the intermediate frequencies and working with light waves.[71]

At this time Townes was working as a consultant to Bell Labs (one day every other week), and there he often talked with Arthur Schawlow, who had been his postdoctoral student at Columbia. Schawlow was also thinking about ways to extend masers to the infrared or even visible range. Working together, they wrote a landmark paper on the conditions required to make masers operate in the infrared, visible, and ultraviolet regions.[72] This paper stimulated a great deal of work, particularly at industrial laboratories, and within three years there were several working lasers (the first being Ted Maiman's at Hughes Research Laboratories).

The maser opened up a new field, which the invention of the laser greatly expanded. In 1959, for the first international conference devoted to the subject, Townes suggested the name 'quantum electronics,' and he edited two books with that name in title.[73] The first issue (January 1963) of the *Proceedings* of the newly formed Institute of Electrical and Electronics Engineers (IEEE) was a 400-page special issue on quantum electronics. From its inception to the present, quantum electronics has been a hybrid field, both physics and engineering. Townes points out that ". . . from the early 1950s to the early 1960s, essentially everyone who contributed to this growing field of masers and lasers . . . were those who had been occupied with the basic science of radio and microwave spectroscopy."[74]

In his years at Columbia University, Townes came to know the radio pioneer Edwin Howard Armstrong, inventor of the regenerative circuit, the superheterodyne receiver, and the super-regenerative circuit. His 1933 patents on frequency modulation (FM) as a way of avoiding static in radio broadcasting were eventually upheld by the courts, but only after his death in 1954.

TOWNES: *I used to see him [Armstrong] at lunch fairly frequently. He was a nice person, interesting. . . . He, at that time, was right in the middle of a patent suit and used to talk a lot about that. He was very annoyed at RCA and other people who wouldn't recognize his patents. Particularly frequency modulation. RCA was claiming that frequency modulation occurs in nature, which then rules out the patentability. Armstrong was arguing it never occurs in nature. That didn't seem quite right to me, but it is not so common in nature. He felt very strongly about this, but he was not too unreasonable. I enjoyed having lunch with him.*

NEBEKER: *He could talk about other things than his patent suits?*

TOWNES: *Yes, he could.*

NEBEKER: *I know that became his main occupation in later years.*

TOWNES: *That's right. He had a lot of patents to fight. By then he'd spent a whale of a lot of money fighting, and he still wasn't winning. But we talked also about general things. He frequently ate lunch with the physicists at the Faculty Club. We'd meet over there. So I would see him with moderate regularity.*

NEBEKER: *He was interested in what was going on in physics generally?*

TOWNES: *I guess he was probably more interested in my work than in others*
           because it had something to do with electronics. He didn't
           particularly talk about high-energy physics or nuclear physics,
           but he talked about electronics. . . .
                   He may have influenced me some in the following respect.
           He eventually jumped off a building and killed himself, and I
           think some of it was family problems. But I'm sure he was very
           depressed about this RCA situation, and he'd spent much of his
           fortune on it.
                   When the maser came along, I was supposed to submit the
           patent case to people in the university so they could patent it,
           which I did. They didn't have any patent policy, so they called
           together a committee to decide what to do. Major Armstrong had
           always been just patenting his own things without going through
           Columbia at all, and he had set a precedent. According to the
           agreement that I had signed for the support of my lab work, the
           patent belonged to Columbia University and I was obliged to
           turn it over to them. They finally decided that they really didn't
           have any policy, and that if I'd like to patent it to go ahead and
           patent it myself. Columbia wouldn't bother about it.
                   So I did that. But knowing Armstrong's problems and how
           difficult these things could be probably was part of my reasoning.
           I don't remember it that clearly. In any case, I decided that I
           didn't want to get locked up in a lot of patent problems and spend
           my life that way. The Research Corporation was handling pa-
           tents and giving the resultant money to universities, and they'd
           given substantial amounts to the physics department. So I gave
           the maser patent to the Research Corporation and let them take
           it over and worry about it and give me a certain fraction back.
           Then I didn't have responsibility for it, and I could forget about
           it. So that's what I did.[75]

Townes remained at Columbia until 1961, though the last two years he
spent almost entirely in Washington as vice president and director of
research of the Institute for Defense Analysis (a nonprofit think tank that
advised the government on weapons systems and on arms control measures). In 1961 he accepted the position of provost and professor of physics
at MIT, and in 1967 he moved to the University of California at Berkeley
as university professor of physics. From 1967 to the present he has given
most of his time to research in astronomy.
    As we have seen, Townes long had an interest in astronomy. Though he
chose not to pursue radio astronomy after the war, he returned to that field
in the late 1950s when he and Columbia associates J. A. Giordmaine and

L. E. Alsop installed a maser on the Naval Research Laboratory's radio telescope in Washington D.C. in 1959 (see Figure 6).[76] (This was the first use of a maser as an amplifier, but maser amplifiers soon found many other applications, notably for satellite transmissions, beginning with Echo in 1960, and in the discovery of the cosmic background radiation by Arno Penzias and Robert W. Wilson in 1965.)

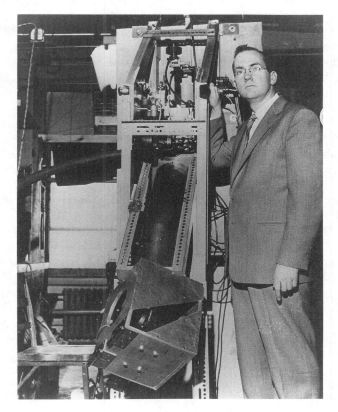

Figure 6. Townes in his laboratory at Columbia University with the first maser amplifier used for radio astronomy.

Townes pioneered the new field of molecular astronomy. In 1955 he had suggested that astronomers look for molecules in interstellar space.[77] Several quite simple free radicals had been discovered much earlier with optical astronomy. Townes's former student, Alan Barrett, made the important discovery of the hydroxyl radical (OH) in 1963. But most astronomers doubted that there were others and that there were any stable molecules, and no one had made the effort to search for them. So when Townes turned to astronomy in 1967, he attempted this and, together with several colleagues, discovered the first stable molecules in interstellar space.[78] Townes wrote recently, "About a hundred different molecules have now been found, including all those thought to be the most important in initiating life, floating in interstellar space and waiting to condense into stars and planets. . . . this new astronomy had its roots in my somewhat arbitrary assignment during the war."[79] (It is interesting that masers have been found in interstellar space, in gas clouds where molecular populations are inverted naturally and amplify microwaves.)

Other areas of astronomy to which Townes has made important contributions are infrared astronomy and, in recent years, infrared interferometry. Borrowing a technique from radio receivers, he introduced heterodyne detection to astronomy by proposing, building, and using an infrared heterodyne detector.[80] The technique was valuable in astronomical spectroscopy because of its high resolution and was especially valuable in astronomical interferometry, the modern form of which depends strongly on the use of lasers in still other ways.

Though the subject of this chapter is the engineering work of Charles Townes, it should be remembered that he has made his greatest contributions as a research physicist and later as an astrophysicist, authoring some 300 papers. In addition, he has been extremely active as a teacher, a university administrator, and a government advisor. At Columbia, MIT, and Berkeley he has directed the work of more than 60 Ph.D. students (including Arno Penzias, Ali Javan, and Elsa Garmire) and almost as many postdoctoral students (including Arthur Schawlow, Reinhart Genzel, and Koichi Shimoda). He was chairman of the physics department at Columbia and provost of MIT. He has been a science advisor to Presidents Eisenhower, Kennedy, Johnson, and Nixon and served on numerous advisory committees for the military, NASA, and the National Academy of Sciences. He has also been active in professional organizations, serving as an editor of half a dozen journals and as president of the American Physical Society.

Townes and Engineering

Yet engineering has been an important element of Townes's career from its inception to the present. Born in Greenville, South Carolina, on 28 July 1915, he regards his engineering experience as beginning with his upbringing on a farm (his father, though, was an attorney): "I think a farm is a good place for both experimental physics and engineering. People have to make do with what's there. They invent things and make things and fix things."[81] As a youth Townes was, by his own report, more interested in nature than in man-made things.[82] He collected leaves, insects, rocks, and bird sightings. He was close to his older brother Henry, who became a biologist, and he went to college expecting to become a scientist, probably a biologist. Nevertheless, he took an interest in gadgets. (See Figure 7.)

A favorite cousin, Frank Dargan, was chairman of the electrical engineering department at Clemson University. He gave Charles and his brother a crystal radio set, which they experimented with. Later, when Charles was in high school, he built his own crystal radio, but never got it working properly. He also had an interest in old clocks and watches, which he used to take apart and reassemble to make them work.

Figure 7.   Charles Townes at about age 12.

Charles's interest in gadgets—and in making inventions—is shown in a letter he wrote, as a 10-year-old, to his older sister Mary Townes, then attending Winthrop College in Rock Hill, South Carolina. (See Figure 8.)

After earning a B.A. and B.S. from Furman University (a small Baptist college) in Greenville in 1935, Townes enrolled at Duke University. The physics department there owned two Van de Graaff generators, but they

Dec. 14, 1925

Dear Mary,

You asked me what I wanted for Christmas. I want mostly hardware so you better buy out a hardware store. I want some tin shears, some money to buy some iron and wood bits, (as I want a particular size I had rather pick out my own) a flat file, a pair of glass cutters, some rifle shot and some one and two penny nails.

I am sorry I have not written you before but I just had so many other things to do I didn't think about it.

Daddy has got a patten [*that is*, patent] thing up that costs a nickel to patten anything. He did it because Henry [Charles's older brother] fusses so much saying I copy him in every thing...

Your brother,

Charlie

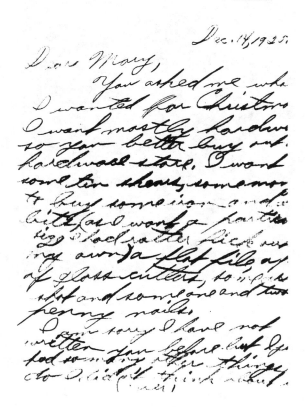

Figure 8. First page of a letter that Townes wrote to his sister Mary on 14 December 1925.

had never functioned very well. Townes undertook to bring them up to a high level of performance by making modifications to the machinery and belts and the way charges were picked off the belts. His master's thesis, which he completed in early 1937, was an analysis of these machines. "I never did any physics with them," he says, so all of this may be regarded as engineering experience.[83]

Townes then moved to the California Institute of Technology as a Ph.D. candidate in physics, but again his thesis work involved a great deal of engineering. He chose W. R. Smythe as thesis advisor: "He was tough, and he didn't have any other students. I felt I would get a lot of attention."[84] There he worked on a gas-diffusion apparatus for isotope separation, one that Dean Wooldridge had built several years earlier, but was not then being used: "... some of the pumps were cracked, and there were other leaks in the vacuum system. A lot of things needed fixing up and improving."[85] (See Figure 9.) He used the apparatus to separate particular isotopes— oxygen 18, nitrogen 15, and carbon 13—and then did spectroscopy on them to determine nuclear spins.[86] Townes's first publication came out of this thesis work; it was not, as one might suppose, a physics paper, but rather a contribution to engineering ("Greaseless vacuum valves").[87]

We have already seen that Townes worked at Bell Labs for more than eight years. His connection with the laboratory, however, began before that day in September 1939, when he first reported for work, going back in fact to his college days in Greenville. He was becoming increasingly interested in physics, but neither his college nor the local library received any physics journals. The local library did, however, receive *Bell System Technical Journal* (AT&T sent the journal gratis), and Townes read it with great interest. A Bell Labs scientist, Carl Darrow, wrote quite a few articles summarizing advances in particular scientific areas. Townes recalls, "To have a summary of a new field of physics written there in a journal was a great opportunity. So I studied those hard."[88] In the 1940s Townes came to be good friends with Darrow.

Townes's connection with Bell Labs did not end when he moved to Columbia in January 1948. For most of his years at Columbia, he was a consultant to Bell Labs, and he has maintained personal relationships with many Bell Labs scientists and engineers to the present. He said recently, "I continue to get help by calling those people. They also, I hope, get a little help and a few suggestions from me from time to time."[89] As testimony to the value of being part of a large scientific and technical community, Townes said, "Frequently, the efficiency of sorting out unfruitful routes for solution of a problem depends on easy access to a friend who has relevant knowledge. Hence, being closely surrounded by other scientific and technical activity is often important to new and exploratory work."[90] His own work shows the fruitfulness of combining ideas from different scientific and technical areas, and he has commented, "What I think really delayed the development of quantum electronics was a lack of the piecing together of ideas from a variety of fields."[91]

Figure 9. Townes in his lab (and office) at Cal Tech in the academic year 1937/38 working on his isotope separation system. On the back of the photo are the words "Looking for leaks—a large portion of most research."

Townes has also maintained contacts with the engineering community in general. Early in his career he joined the Institute of Radio Engineers (one of the predecessor societies of the IEEE) because he valued its journal, *Proceedings of the IRE*. (It is indicative of the origins of radio astronomy that many of the seminal papers in that field, including Townes's 1959 article on the maser amplifier for radio astronomy, appeared in *Proceedings of the IRE*.)[92] In 1958 Townes was awarded the Morris N. Liebmann award of the IRE, and in 1961 the David Sarnoff Award of the American Institute of Electrical Engineers (the other predecessor society of the IEEE). In 1967 he received the highest award of the IEEE, the Medal of Honor, and in 1988 he was named an IEEE Life Fellow.

The traditional view is that technology is derivative of science: Scientists make discoveries and gain understanding of how things work, and engineers later apply scientific knowledge to practical problems. Historians of technology have emphasized the influence in the other direction: Technology both permits scientists to answer questions and raises new questions.[93] Townes has, in numerous writings, expounded on the symbiosis of science and technology: "While scientists are rather familiar with the flow of contributions from basic to applied work, and the phenomenon is recognized in generality by the broader public, I do not believe adequate recognition is given to the reverse flow—the contribution of technology to basic science. Convenient and sophisticated instrumentation is but one example. More broadly, such contributions encompass important scientific discoveries produced by applied research, the development of industrial and commercial products on which much basic research depends, and new technical possibilities that emerge from applied work in industry and in military and space programs."[94] He has also commented that, "Science and technology really have to develop together. That's a great disadvantage for countries that don't have a good technological base. Their scientists are at a considerable disadvantage."[95]

As shorthand one may say that wartime radar systems led to microwave spectroscopy, and that microwave spectroscopy led to masers and lasers. This is shorthand because it was, of course, not technology spawning technology but people discovering and inventing, people with particular talents, motivations, and knowledge. The story of Charles Townes shows how fruitful a combination of scientific and engineering knowledge can be, and it reveals also several important aspects of the modern relationship between science and technology: the scientist's ability to contribute to the advance of technology, the engineer's ability to contribute to the advance of science, and the personal and institutional connections between industrial R & D and academic science.

---

[1] The information about Charles Townes contained in this article comes mainly from the following sources: (1) an extensive oral history interview of Townes conducted by the author 14 and 15 September 1992 (available at the Center for the History of Electrical Engineering); (2) the laboratory notebooks kept by Townes in his years at Bell Labs and papers from the files of the projects on which he worked (available at the AT&T Archives); (3) interviews of Townes by William V. Smith (18 to 20 June 1979), Joan Lisa Bromberg (28 and 31 January 1984), and Finn Aaserud (20 and 21 May 1987), all of which are available at the archives of the Center for History of Physics of the American Institute of Physics; (4) a 35-hour oral history interview conducted by Suzanne B. Riess, 5 November 1991 to 3 July 1992, for the Regional Oral History Office (ROHO), The Bancroft Library, University of California, Berkeley (the interview will be available in final manuscript form in 1994); and (5) published writings by Townes and others.

[2] See "The early history of microwaves," Chapter 8 of Henry E. Guerlac's

Radar in World War II (Tomash Publishers and American Institute of Physics, 1987), pp. 185–240; and "The background to the development of the cavity magnetron" by R. W. Burns in a book he edited called *Radar Development to 1945* (Exeter: Peter Peregrinus, 1988).

[3] Interview 1992, p. 10

[4] The final sentence of the preface to the book, *Static and Dynamic Electricity* (New York: McGraw-Hill, 1939), reads, "In particular, he [the author] wishes to thank Dr. Charles H. Townes for checking all the derivations in the final manuscript and the answers to those problems not appearing in the preliminary edition."

[5] Interview 1992, p. 116.

[6] Interview 1992, pp. 116–117.

[7] Bell Labs Notebook 17015 (1939–1941), p. 7. The device is further described in the six following pages. Townes wrote up some of this work as a Bells Labs Technical Memorandum, "Energy relations in electron flow through non-uniform high frequency fields," 2 January 1940.

[8] Interview 1992, pp. 8, 11–12.

[9] Patents 2,438,954 (electronic oscillator of the cavity resonator type); 2,439,381 (computing bombsight, with Darlington and Wooldridge); 2,457,287 (air-speed indicating system); 2,488,448 (computing circuit for determining bomb release points, with Wooldridge); 2,511,197 (averaging device, with Darlington and Wooldridge); 2,544,754 (electron camera tube); 2,701,098 (concealed ground target computer); 2,707,231 (frequency stabilization of oscillators); 2,707,235 (frequency selective systems); and 2,819,450 (molecular resonance modulators and demodulators).

[10] Interview 1992, p. 9.

[11] The basic maser-laser patent is number 2,879,439 (issued 24 March 1959), and the basic laser patent (with Schawlow) is 2,929,922 (issued 22 March 1960). The only patent Townes has since received is number 3,469,107 for stimulated Brillouin parametric devices (with Stoicheff, Garmire, and Chiao), issued 23 September 1969.

[12] Bell Labs Notebook 17015 (1939–1941), pp. 21–46.

[13] Interview 1992, pp. 17–18.

[14] *IEEE Transactions on Magnetics*, December 1969.

[15] Townes did not make many notebook entries in the month and a half he was in Bozorth's group [Bell Labs Notebook 17015 (1939–1941), pp. 48–56]; he reports that in this time he was "mainly learning" (Interview 1992, p. 21).

[16] Bell Labs Notebook 17015 (1939–1941), pp. 57–127. Work recorded on pages 57 to 120 is classified as Case 35887; work on pages 121 to 127 as Case 37641.

[17] Bell Labs Technical Memorandum 6 January 1941, "Theory of cathode sputtering in low voltage glow discharges"; Townes, "Theory of cathode sputtering in low voltage gaseous discharges," *Physical Review*, vol. 65 (1944), pp. 319–327.

[18] Interview 1992, p. 28.

[19] S. Millman, ed., *A History of Engineering and Science in the Bell System: Communications Sciences (1925–1980)* (AT&T Bell Laboratories, 1984), pp. 356–359.

[20] Interview 1992, pp. 34–35.

[21] M. D. Fagen, ed., *A History of Engineering and Science in the Bell System: National Service in War and Peace (1925–1975)* (Bell Telephone Laboratories, 1978), pp. 163–170.

[22] Ibid., pp. 134–136.

[23] Interview 1992, p. 35.

[24] Interview 1992, pp. 35–36.

[25] Interview 1992, p. 29. Another of Kelly's initiatives was directed toward all Bell Labs employees: Communications Development Training, usually called Kelly College, was a large in-house training program for the continuing education of company employees.

[26] ROHO Interview, p. 247.

[27] See the first two chapters of Fagen, *A History of Engineering and Science.* After the war, the number of military projects decreased rapidly; in 1947 they accounted for 15 percent of the Bell Labs budget.

[28] Interview 1992, p. 38.

[29] Interview 1987, p. 27.

[30] Bell Labs Notebook 17015 (1939–1941), p. 125.

[31] This was the SCR-519 radar combined with a version of the M-9 computer. See Fagen, *A History of Engineering and Science*, pp. 91–93.

[32] Bell Labs Notebook 17870 (1941–1942), pp. 1–200.

[33] Bell Labs Notebook 17870 (1941–1942), pp. 118–119.

[34] Interview 1992, pp. 41–43.

[35] Interview 1992, pp. 63–64.

[36] Interview 1992, p. 71.

[37] See "Proposed Bomb Sight: Memorandum for file," 28 March 1942, Bell Labs Archives, Case 23665.

[38] See "Conference Notes —Case 23665: Memorandum for file," 29 October 1942, Bell Labs Archives, Case 23665, and Bell Labs Notebook 17870 (1941–1942), pp. 138–156.

[39] Bell Labs Notebook 19729 (1944–1946), p. 7.

[40] Bell Labs Notebook 19729 (1944–1946), pp. 12–13, and Interview 1992, p. 84.

[41] Interview 1992, p. 50.

[42] Bell Labs Notebook 18774 (1942–1943), p. 101.

[43] Interview 1992, p. 86.

[44] Interview 1992, p. 86.

[45] Fagen, *A History of Engineering and Science*, p. 9.

[46] Karl Jansky, "Electrical disturbances apparently of extraterrestrial origin," *Proceedings of the IRE*, vol. 21, 1933, pp. 1387–1398.

[47] Townes, "Interpretation of radio radiation from the Milky Way," *Astrophysics Journal*, vol. 105, 1946, pp. 235–240.

[48] Fagen, *A History of Engineering and Science*, pp. 89–113.

[49] ROHO Interview, p. 169.

[50] Interview 1992, p. 58.

[51] Interview 1992, p. 75.

[52] Bell Labs Notebook 19729 (1944–1946), pp. 96–101.

[53] Bell Labs Notebook 19729 (1944–1946), pp. 106–108.

[54] [Harvard] Radiation Laboratory Report No. 43-2, "The atmospheric absorption of microwaves," 27 April 1942.

[55] Interview of Townes, *Laser Pioneer Interviews* (Torrance, CA: High Tech Publications, 1985), pp. 35–47.

[56] Interview 1992, pp. 57–58.

[57] See ROHO Interview, pp. 207–209, and Interview 1992, pp. 108–110.

[58] See chapter 2 of J. S. Hey's *The Evolution of Radio Astronomy* (New York: Science History Publications, 1973).

[59] Memorandum by Townes entitled "Applications of microwave spectroscopy" (undated, probably May 1945), Bell Labs Archives.

[60] Bell Labs Notebook 20464 (1945–1946), p. 2.

[61] New York: McGraw-Hill, 1955.

[62] C. H. Townes and A. L. Schawlow, *Microwave Spectroscopy* (New York: McGraw-Hill, 1955), p. 1.

[63] ROHO Interview, p. 268.

[64] Bell Labs Notebook 21268 (1946–1947), pp. 142–144, and Patent 2,544,754 (electron camera tube).

[65] Interview 1992, p. 118.

[66] Four people in industrial research labs with experience in radar technology (besides Townes, they were W. E. Goode at Westinghouse, A. H. Sharbaugh at GE, and W. D. Hershberger at RCA) began to explore the new field of molecular spectroscopy after the war. All were permitted to do so for some time (showing the freedom scientists in industry had), but because the field seemed not to have much application, they were not given much support nor, in some cases, allowed to continue work in that area.

[67] Interview of Townes in *The Way of the Scientist: New Views from the World of Science and Technology* (New York: Simon and Schuster, 1966), p. 61.

[68] Paul Forman, "Inventing the maser in postwar America," *Osiris*, 2nd series, vol. 7, 1992, pp. 105–134.

[69] Townes, "The laser's roots: Townes recalls the early days," *Laser Focus*, August 1978, pp. 52–58.

[70] Interview 1979, p. 62.

[71] Townes, "The laser's roots: Townes recalls the early days," *Laser Focus*, August 1978, pp. 52–58.

[72] Townes and Schawlow, "Infrared and optical masers," *Physical Review*, vol. 112, 1958, pp. 1940–1949.

[73] *Quantum Electronics* (1960) and *Quantum Electronics and Coherent Light* (1965).

[74] Townes, "The laser's roots: Townes recalls the early days," *Laser Focus*, August 1978, pp. 52–58.

[75] Interview 1992, pp. 126–128.

[76] J. A. Giordmaine, L. E. Alsop, C. H. Mayer, and C. H. Townes, "A maser amplifier for radio astronomy at X-band," *Proceedings of the IRE*, vol. 47, 1959, pp. 1062–1070.

[77] Townes, "Microwave and radio-frequency resonance lines of interest in radio astronomy," in H. S. van de Hulst, ed., *IAU Symposium No. 4: Radio Astronomy* (Cambridge UK: Cambridge University Press, 1957), p. 92 [the talk was given in August 1955].

[78] Townes, A. C. Cheung, D. M. Rank, D. D. Thornton, and W. J. Welch, "Detection of NH_3 molecules in the interstellar medium by their microwave emission," *Physical Review Letters*, vol. 21, 1968, pp. 1701–1705.

[79] Townes, "Reflections on my life as a physicist," *Bulletin of the Center for Theology and the Natural Sciences*, vol. 12, no. 3, 1992, pp. 1–7.

[80] The idea was initially advanced in Townes and H. A. Smith, "Frequency conversion and detection of infrared radiation," *Polarisation matière et rayonnement* (Presses Universitaires de France), pp. 467–483. The first paper giving results obtained using the heterodyne detector was Townes, M. A. Johnson, and A. L. Betz, "10-mm heterodyne stellar interferometer," *Physical Review Letters*, vol. 33, 1974, pp. 1617–1620. Townes reports that he seldom writes an instrumentation paper, waiting to describe the instrument until he has obtained scientific results with it.

[81] Interview 1992, p. 114.

[82] Townes, "Reflections on my life as a physicist."

[83] Interview 1992, p. 113.

[84] Interview 1992, p. 114.

[85] Interview 1992, p. 114.

[86] ROHO Interview, p. 117.

[87] *Review of Scientific Instruments*, vol. 9, 1938, pp. 428–429.

[88] Interview 1992, p. 108.

[89] Interview 1992, p. 101.

[90] Townes, "Science, technology, and invention: Their progress and interactions," *Proceedings of the National Academy of Sciences of the USA*, vol. 80, 1983, pp. 7679–7683.

[91] Townes, "Quantum electronics at Columbia University," *Proceedings of the Fortieth Anniversary Symposium of the Joint Services Electronics Program (JSEP)* (U.S. Army Research Office, 1987), pp. 71–88. In this paper Townes tells of adapting an idea he got from a biologist colleague at Columbia to understanding noise in a maser amplifier.

[92] See the list of papers on pages 190 to 204 of Hey, *The Evolution of Radio Astronomy*.

[93] For example, Thomas Kuhn, in "Energy conservation as an example of simultaneous discovery" (in Marshall Clagett, ed., *Critical Problems in the History of Science* (Madison: University of Wisconsin Press, 1959), pp. 321–356), shows how important knowledge of the steam engine was in the formulation of the principle of conservation of energy.

[94] Townes, "Science, technology, and invention: Their progress and interactions," *Proceedings of the National Academy of Sciences of the USA*, vol. 80, 1983, pp. 7679–7683. More than a hundred years earlier, in 1874, another

physicist who was much involved with engineering, William Thomson (later Lord Kelvin), made a similar point: " . . . I have more to say respecting the reflected benefits which electrical science gains from its practical applications in the electric telegraph than of the value of theory in directing, and aiding, and interesting the operators in every department of the work of the electric telegraph," *Popular Lectures and Addresses*, Vol. II (London: Macmillan, 1894), p. 211.

[95] Interview 1992, p. 104.

CHAPTER 4

GORDON TEAL

Figure 1. Gordon Teal, born 10 January 1907, is a recipient of the IEEE Medal of Honor, a member of the National Academy of Engineering, and a current resident of Dallas, Texas.

Finding
the Right Material
Gordon Teal
as Inventor and Manager

In 1930 Robert Williams, the head of the chemistry department at Bell Telephone Laboratories, spelled out the role of chemistry research for AT&T: "In the Laboratories chemists act chiefly as advisors and critics. They concern themselves with such problems as the theory of chemical structure as related to dielectric properties and simultaneously attack the task of making an improved substitute for gutta-percha which renders possible a transatlantic telephone."[1] He went on to identify some of the other outstanding research projects, all of which were devoted to enhancing the durability of telephone equipment. He closed his list with a discussion of finishes for telephones and measures to extend the life of telephone poles.

In fact, chemistry was more relevant to electrical engineering than Williams suggests. Michael Faraday's interest in the subject, as well as that of Thomas Edison, speaks to the fundamental link between the two fields. Graduates of the first electrical engineering bachelors degree programs, such as the one at the University of Wisconsin designed in the late nineteenth century by noted EE educator Dugald Jackson, were required to complete one year of rigorous chemistry instruction. The early success of General Electric at the beginning of this century was due in no small part to chemical breakthroughs in filament manufacture. At Bell

Labs, however, where electrical engineering and physics were the principle activities, chemistry had only marginal status in 1930.

Thirty-two years later, scientist Richard L. Petritz, in an article entitled "Contribution of Materials Technology to Semiconductor Devices," pointed out that materials technology, the burgeoning discipline that had evolved from chemistry, was one of the basic disciplines underlying the design and development of solid-state electron devices.[2] Inasmuch as those devices were already well on their way to revolutionizing electrical engineering, Petritz's observation indicates that the relation of materials science to electrical engineering had changed dramatically from the time Williams wrote the passage above. That transformation, one of the most important in electrical engineering history, is well illustrated by the work of one of its central figures, Gordon Teal. Teal distinguished himself as the inventor of the first processes to grow germanium and silicon junction transistors, and also as a leader of scientific research and manager of science in the industrial context. His career exemplifies many important aspects of the research scientist's role in technological development in the mid twentieth century.

Early Days

Gordon Kidd Teal was born on 10 January 1907 in South Dallas, Texas. His father, Olin Allison Teal, had come to Texas in 1897 from Georgia, ready to make a go of running a five-and-dime store that Olin's uncle had established sometime earlier. The two men did well, and soon a "Duke and Teal's" could be seen in many Texas towns. Olin left that business in 1910 to begin a real estate enterprise, which also thrived. The Teals prospered, at least until the Great Depression, and the family lived free from want.[3]

The Teals lived in a simple community and were dedicated members of the First Baptist Church of Dallas. Gordon's father, who became deacon in 1905, brought his three children—Gordon, his older sister, and his younger brother—to church every week. Young Gordon accepted the church's teachings and became a faithful Baptist.

Olin Teal provided an academically stimulating atmosphere for his children. Born into humble circumstances and receiving only home instruction prior to attending high school, Olin nevertheless managed to graduate from college, distinguishing himself with high marks in Latin, French, and mathematics. He supported himself as a substitute school teacher and completed his courses in only two years and a summer session. He was not particularly scientifically inclined, but Olin's example set a standard of excellence for his children to emulate. Gordon's mother, Azelia, reinforced that message by assuring her son that he was especially capable. With hard work, she promised him, he could live up to the model of his father.

She encouraged his boyhood interest in science, and when he was in high school, told him he had the gifts to be the best at it. Teal, inspired by his mother's praise, applied himself to exceed her expectations. He graduated in 1924 from the Bryan Street public high school as valedictorian, having earned the highest marks of any student in Dallas.

With his successful high school experience, Teal hoped to attend the most prestigious school he could. He looked north, to the Massachusetts Institute of Technology. His mother, however, entreated him to remain close to home and attend a Baptist school. He agreed to try Baylor University for one year. The school appealed to Teal, and he met a woman there, an outstanding student named Lyda Smith whom he was eager to stay close to. So, without complaint, Teal agreed to finish his education at Baylor.

At Baylor, Teal concentrated on math and chemistry, two subjects he both enjoyed and excelled in. Engineering was not offered at Baylor, and Teal elected not to take physics courses so that he might better focus on his chosen fields. He studied assiduously and achieved high marks. Academic dedication did not prevent him from exploring a variety of pastimes, however. He led the Latin club and the Baylor chapter of the Kappa Epsilon Alpha Honor Scholarship Society as president, and served as vice president of his senior class. He ran on the track team and took great pleasure in being a member of the Baylor Chamber of Commerce. In 1927 Teal graduated from Baylor with a double major in math and chemistry.

MIT still tempted Teal, and in his last years at Baylor he began planning to continue his studies there. Split between his twin interests of math and chemistry, he finally selected the latter as his preferred subject. The tangibility of chemistry held greater appeal to his practical sensibilities than did math. One of his chemistry professors at Baylor recommended that Teal consider Brown University over MIT. Teal investigated, and developed a positive feeling about the program there after meeting the head of the department, Dr. Charles A. Kraus. Brown offered Teal a full scholarship to attend, and Teal, who was eager to spare his family the expense of additional schooling, again put off MIT. In the fall of 1927, he headed to Providence, Rhode Island, to begin his graduate work in chemistry as an Edger L. Marston Scholar.

The late 1920s was a dynamic time for chemistry. Advances in quantum physics offered a radically new picture of atomic processes, and chemists had an opportunity to use these to deepen their understanding of familiar atomic processes.[4] When Teal arrived at Kraus's lab at Brown, however, he found many of the students there were instead working along more traditional lines to discover the basic chemical properties of the esoteric element germanium. Teal had little experience with this metal—indeed, few outside of Kraus's lab knew very much about it—and it quickly captured his imagination. Possessed of "a strong desire to exploit discoveries for practical use," Teal was challenged by the apparent uselessness of germanium.[5] In 1931, he finished his dissertation research on certain reactions of germanium, mixing sodium germanyl and potassium germanyl with alkyl halides to prepare the methyl-, ethyl-, and propyl-germanes that standard theory predicted. The reactions Teal studied were untried cases of a well-known general reaction.[6] He explored them in an effort to find a use for a material to which he had formed an attachment that is best described as esthetic.

. . . I always have had considerable curiosity, with particular interest in new things and new ideas. This has led me over the years to a closer look at some of the basic phenomena in science and stimulated a strong desire to exploit discoveries for practical use. I also share the delight in the esthetic . . . I enjoy adventure and, in particular, I am interested in pioneering, that is, the kind of pioneering demanded by our rapidly changing and highly technical society. These interests have greatly influenced my decisions and were often crucial in the directions my career has taken.[7]

Even before Teal submitted his dissertation, he began his professional career as a chemist. In 1929, he toured Bell Laboratories, eager to investigate operations inside the esteemed facility. Teal and Bell were mutually impressed. Asked not to start work anywhere before talking to Bell, Teal ceased flirting with the idea of working at Du Pont or some other laboratory. When Bell made him an offer in the summer of 1930, he accepted. Overcoming reservations he had about living in New York City (the laboratory was located in the city, in a structure on West Street, at that time), Teal began his 22-year career at Bell Laboratories in August 1930.

Teal on TV

Almost immediately after Teal joined Bell Labs, the Depression forced AT&T, the Lab's corporate parent, to institute a hiring freeze on new scientific staff. The Lab reduced its payroll expenses by allowing the attrition of its workforce, laying off some staff, and reducing the hours of those that remained. Teal, who married Lyda on 7 March 1931, experienced deep concern about his future at Bell, but he was reassured that his low seniority posed no threat to his job security. He occupied the time that the reduced hours program had liberated for him with study of heavy hydrogen in collaboration with future Nobel laureate Dr. Harold Urey, who was at Columbia University.[8]

Shortly after establishing himself in the chemistry department, Teal began looking at ways to purify mercuric oxide at the request of Bell Labs' television division. Teal's chemical abilities had attracted the attention of Herbert E. Ives, the leader of that division. Television tubes demanded skilled chemists to prepare the light-sensitive substances and glass for the tubes. By Teal's second year at Bell, he was transferred from chemistry to television.

Bell Labs began television research in 1925 and soon became the preeminent research organization in the field. In April 1927 Ives and his colleagues transmitted pictures of a speech given in Washington, D.C., by Secretary of Commerce Herbert Hoover to an audience in New York through coaxial cable. They also broadcast the same scene over the air the considerably shorter distance between Whippany, New Jersey, and New York City. The sharpness of Bell's pictures impressed those who witnessed the demonstration. No better picture would be produced by any system for years to come.[9]

Bell's early success was achieved with a mechanical television system. In it, the camera apparatus scanned the image by the spinning action of an internal disc. Ives favored this approach to its alternative, electronic scanning of the image by a cathode ray within the camera tube. Ives experimented with an electronic system in 1926 and the results convinced him that an electronic system posed too many serious problems. Thereafter, he focused Bell's efforts to mechanical systems, rejecting the electronic idea again when it was proposed by a member of his staff in 1930.[10] By 1933, however, the announcement of an electronic scanning tube called the iconoscope by RCA engineer Vladimir Zworykin prompted Ives to soften his position. Before 1933 ended, Bell had begun its own research on iconoscope-like tubes. It was at this stage that Teal joined the television effort.

Teal worked principally on a critical element within the iconoscope, the light-sensitive target called the mosaic. The tube worked by focusing light from an object onto the mosaic and then converting the physical reaction there into an electrical signal by bombarding the mosaic with a stream of electrons from a hot cathode. The mosaic posed several difficult research challenges. Chemists needed to develop a substance that was highly sensitive to light, and they also had to develop a technique to apply it to the mosaic substrate. No less an authority than Zworykin himself conceded that "The most difficult item of construction in the image multiplier iconoscope is the mosaic."[11]

Teal's work on a variety of mosaic types exemplifies the significance of patient, technique-oriented innovation to his scientific investigations. He began in 1934 by experimenting with compounds of silver and cesium or potassium to create a substance, described as photoemissive, which reacted to light by ejecting electrons. This was the most common approach to mosaic fabrication, but Teal did not work with it long. As early as March 1935, he was exploring photoconductive mosaics, which generated voltage between illuminated spots and nearby dark spots. Later, he concentrated his efforts on still a different type of mosaic, one that responded not to incident light, but rather to electrons that were generated by light from the object focused on a photocathode and then directed toward the mosaic.[12]

One challenge with these latter mosaics was to fix the sensing material into the substrate in a regular and controlled way. Among the techniques that Teal developed was to cover a magnetic pole piece with a sticky substance and then drop tiny magnetic wires onto the piece. The magnetism of the pole piece caused the wires to stand up straight. He caught the wires' top ends in a support and then removed the sticky substance from their bottoms. He coated the wires with insulator and then filled in the space between the insulated wires with a conductor. After releasing the tops of the wires from their support and electroplating, he was left with the desired configuration of sensitive material, conductors, and insulators.[13]

Another method Teal invented was to take a metallic screen with a very fine mesh (400 holes to the linear inch) and coat the inside wall of the apertures with insulator. He then laid the screen on a porous material and poured over it a fluid that contained small particles of photosensitive material. The particles were small enough to fall through the apertures, but too large to pass through the pores of the object upon which the screen rested. With the particles each inside an aperture, lying in their pools of fluid, he used a pressure difference to force the fluid through the porous substance, leaving the particles alone to fill the apertures.[14]

One other approach to mosaic design quickened Teal's interest. W. H. Hickok, a researcher at RCA, was taking advantage of the thermoelectric properties of germanium, Teal's "pet" element, to build a pick-up that was sensitive to far-infrared radiation.[15] The germanium mosaic reacted not to the light reflected off objects but to their temperature. RCA imagined a scope that could aid navigation by televising objects through fog or other visual interferences. Teal has recalled trying to interest his superiors at Bell Labs in germanium research in the 1930s.[16] Although there is no record of what specifically he proposed, it appears probable that he was suggesting following up on the RCA experiments. Besides having a latent interest in the element itself, Teal was eager to match the accomplishments of other television research laboratories. The keen sense of competition that drove him is evidenced by his practice of pasting the press clippings of other labs' successes into his lab notebook for inspiration.[17]

Bell Labs did not share Teal's enthusiasm for general television research. From the high point of Ives's triumphant 1927 demonstration, AT&T watched Bell's leadership in television research slip away during the 1930s without much panic or concern. In 1925, AT&T president Walter Gifford decided to pull his company out of most businesses not related to telephone service in the United States. Activity in television, along with other areas such as radio broadcasting, household appliances, and foreign telephone service, declined at AT&T. The television research that continued considered TV primarily as a visual extension of the telephone— allowing pictures to accompany words in point-to-point communications.[18] Although Bell continued to spend an average of $133,000 per year on television between 1932 and 1937, and then $335,000 per year between 1938 and 1940, their commitment remained far below that of RCA, for instance, which spent upwards of $50 million developing a working television system.[19] In deciding how to spend these comparatively modest allocations, Bell's management applied their most important criterion— the relevance of the proposed research to AT&T's principal business of providing telephone service. It is little surprise that in this climate, Teal's proposal for germanium mosaics (sensitive to light outside of the visible spectrum) met with rejection.

Teal got so involved in his television work that he moved beyond the boundaries of his expertise in chemistry. Without any academic training in electrical engineering, Teal began designing electrical components and systems with a proficiency in EE he acquired solely through experience in the lab.[20] He patented several electron multipliers, assemblies of electrodes that amplified signals by augmenting beams of electrons. This work contributed to his eventual composition of an entire system for television, which he patented in 1941. Teal's system, following the "flying spot" innovation that made Ives's television transmission for Bell a success in 1927, moved a sheet of light across the subject, successively illuminating parallel linear sections of it. He also worked on a tube that could reduce the number of frames per second needed by storing images on the mosaic (as the iconoscope did) and scanning at quick rates (what would today be called "faster than real-time") in order to cut the bandwidth needed to transmit an image.[21] He worked on these ideas, and shared many of his patents, with the small group of men he directed, including A. W. Treptow, R. L. Rulison, and B. A. Diggory.

Semiconductor Days . . . and Nights

Television research at Bell ended abruptly with the US entry into World War II and much of the electronic expertise in the department was reassigned to radar systems.[22] Teal, a chemist foremost, despite his work with electronics, was set to work developing materials crucial for military equipment. Teal moved suddenly from engineering complete television systems to investigating molybdenum carbide coatings for gun barrels, dies, and rocket nozzles. If the change to a less glamorous assignment saddened him, his feelings are largely hidden behind the stiff formality of his 10 February 1942 notebook entry: "On Friday morning, Feb 6, 1942, Mr. S. O. Morgan [Stanley Morgan, who later headed Bell's chemistry department] informed me that I am to work on the silicon detector problem in collaboration with Mr. Shive."[23]

Any remorse Teal felt over the end of television work was balanced by the excitement of the new opportunity that war-related research offered him to work with semiconductors. The war stimulated an unprecedented interest in those metals of which Teal was so fond. Silicon, germanium's sister element on the periodic chart, proved particularly valuable for radar systems. AT&T began using the material to make rectifiers, components that turn alternating current into direct current, in early 1942. By 1944, Western Electric, the manufacturing division of AT&T, was turning out over 50,000 units per month.[24]

Teal saw a chance, once again, to try to interest his superiors in research on germanium. He hoped that rectifiers made of germanium would outperform the silicon rectifiers that Bell was already exploring. Early in 1942 he fabricated some germanium rectifiers, the first germanium devices made at Bell Labs, and in June, the director of research at Bell Labs, Mervin J. Kelly, authorized Teal to continue his work.[25] Within a few months, however, Teal found his laboratory assistance for the project disappearing. Sensing that he was being discouraged from germanium research, Teal dropped the project and turned to radar attenuators. Soon after, his germanium rectifier program was revived under a contract from the MIT Radiation Laboratory, but Teal was unable to participate in the work because he had contracted pneumonia in a work-related accident. The germanium research proceeded without him.

Teal spent much of the war studying attenuators, the elements in microwave systems, such as radar, that function as resistors do in electric circuits. Attenuators control the power of microwave signals traveling within the wave guides that make up microwave circuits, allowing operators greater control of their systems. Teal experimented with ceramics, plastics, films, and rubbers in an effort to optimize properties such as power handling, sensitivity to temperature, moisture absorption, heat conductivity, energy reflectivity, and physical size.[26] The work involved consideration of the physical properties of the materials under study and the effect of the attenuator's shape on its performance.

The war disrupted the course of Teal's career, but his new research areas represented only a temporary interlude, not a permanent change. The tasks he undertook were selected more on the basis of wartime expediency than personal interest, and he did not directly continue any of them after Bell demobilized. Still, some of his war work helped prepare him for the next, most important, phase of his career.

After the war, Bell took advantage of the familiarity that Teal had gained with silicon by assigning him responsibility to manage the materials development for two new, inexpensive, silicon-carbide varistors. The varistor, an electrical component that changes resistance when the applied voltage changes, is an essential element in the telephone handset. Western Electric planned to manufacture four to eight million varistors per year. The design requirements included close tolerance (±10 percent), 20-year reliability, and low cost (about five cents per varistor.) The assignment had satisfactions for Teal. It was an important project—Western Electric would come to manufacture over 100,000,000 varistors—and he was able not only to contribute to the materials side, but also to work closely with electrical engineers on the device's design. Judged within the context of AT&T's priorities, the job was as significant as any other at Bell.[27] At the same time, however, developments elsewhere in the lab would steal Teal's most ardent interest away from the varistor project. In 1948 Teal heard that germanium was being used in a new solid-state amplifier called the transistor.

By 1948 transistor research already had a substantial history at Bell Labs. The development effort began with the reign of Mervin Kelly as director of research. Kelly had directed the vacuum tube division between 1928 and 1934 and was sensitive to the limitations of the components that AT&T relied upon to switch and amplify the electrical signals that traveled along its phone lines. The company used relays to switch calls from one line to another, but relays worked slowly and were prone to fail when dirt accumulated on their metal contacts. Kelly sought an electronic replacement for the electromechanical relays, but electronic devices were not free of problems either. Tubes were fragile and expensive. They generated heat while doing nothing but standing ready. This wasted power and shortened life span, which diminished the tubes' reliability. When Kelly was promoted to director of research in 1936, he initiated measures to exploit the nonlinear electrical properties of semiconductors as an alternative to tubes and relays.

Kelly's research program was interrupted by the war, but even before the fighting ended, he began to reorganize a solid state amplifier research effort at Bell Labs. He created a multidisciplinary group, which he placed under the charge of theoretical physicist William Shockley and chemist Stanley Morgan. Shockley began by focusing his team's efforts on silicon and germanium, two semiconducting elements that had become quite familiar to scientists, if not Shockley himself, through wartime research. Of the two, germanium was the simpler material. Working steadily with germanium, two members of Shockley's team, Walter Brattain and John Bardeen, built a working amplifier on 23 December 1947. It was soon after named the "transistor" and word about the breakthrough spread throughout the laboratory.[28]

Teal recognized the enormous potential of the solid state amplifier when he first heard about it informally in early 1948. Intrigued by the germanium composition of the device, he resolved not to let this rewarding research opportunity slip by him, as the germanium rectifier project had. He pressed for a new line of germanium research, stressing in several memos to research management that a deep understanding of, and control over, the properties of germanium were crucial to satisfactory transistor performance.[29] He hoped to improve the germanium refining techniques and, more importantly, to grow single crystals of germanium to use for transistors as a replacement for the polycrystalline samples that the Lab was using. He argued that just as superior vacuum resulted in improved performance of vacuum tubes, so would transistors benefit from purer, more uniform germanium. Teal likened the unwanted impurities in germanium and the grain boundaries of the multiple crystals to poor vacuums in tubes.[30]

Teal's proposals were greeted with indifference. As with the germanium in the iconoscope mosaics episode, Bell management did not agree with Teal about the value of the research he proposed. In the case of crystals for transistors, however, it was not a matter of intriguing research losing to the pragmatic need for useful products. Teal promoted the single crystals from the most practical standpoint, offering improved predictability of transistor behavior. Bell management disagreed, but on a strictly scientific basis. Shockley believed that the germanium the lab used was adequate and that single crystal germanium would offer no advantage.[31] Bell's research management, accepting Shockley's opinion, regarded Teal's offer of single crystal germanium as irrelevant to the transistor project and refused to support it.

In the pursuit of new knowledge bearing on rather deep questions and useful concepts of natural phenomena, the scientist has learned that certain attitudes are helpful. He has learned to suspend his prejudices, to stop and take a fresh look. He finds it a helpful exercise to decide what are the relevant and the irrelevant facts. He is interested in an economy of conceptual symbols and relations in which to express his observations. He has found in this economy an expression of principles that are often a more permanent part of accepted science than are the facts which gave rise to the principles. He has found it useful to make precise measurements under known conditions in order to check the correctness of his conjectures and to add precision and clarity to his views. He finds it a good habit to be logical but also, paradoxically, personal knowledge of major advances made in science in which logic was not the dominant feature has made him an enemy of categorical imperatives and authoritarianism in views. He is relentless, dissatisfied with current views and driven by a fierce curiosity.[32]

Teal soon found a chance to grow single crystals of germanium, however. In late September 1948, Teal learned that a colleague of his, John Little, needed a thin rod of germanium fabricated. Teal offered to prepare it, adding as a bonus that the germanium would be a single crystal. By December Teal and Little succeeded in creating a monocrystal rod of germanium by adapting a technique developed in 1918 known as the Czochralski method. (See Figure 2.) They dropped a small seed of germanium into a molten pool of the purified element. As they slowly withdrew the seed, the liquid germanium, called the "melt," solidified uniformly, extending the crystal of germanium along the pattern established by the seed. Teal showed his crystals to Jack Morton, who directed the transistor development effort. Morton was interested in what Teal was doing, but would not take Teal off the important varistor project, nor would he provide Teal with extra laboratory space. Morton did agree to purchase equipment for Teal to continue his experiments and allowed him to use the metallurgical shop after normal working hours. With that opportunity, Teal began a maverick program of "bootleg" research.[33] He would roll his crystal pulling equipment into the lab in the late afternoon as the staff there prepared to depart and would experiment when the room was unoccupied. When he would finish working, usually at two or three in the morning, he had to disconnect his machinery and stow it before the metallurgical staff arrived at work the following morning. For most of 1949, Teal occupied his nights and weekends growing progressively superior crystals.

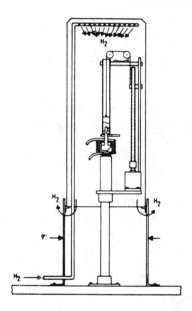

Figure 2. Schematic drawing of the machine designed by Teal and Little to grow monocrystals of geranium.

This uncomfortable situation improved when it was recognized that Teal's work might have some bearing on the progress of a different transistor project managed by Shockley. Soon after Brattain and Bardeen's breakthrough, Shockley began theoretical work on a new type of amplifying semiconductor device, which he called the "junction transistor." This transistor differed from the first sort in that it exploited semiconducting properties in the bulk of its germanium base. Bardeen and Brattain's transistor was active only at the points where metal leads touched the semiconducting crystal.

Any semiconductor can be one of two sorts, p-type or n-type. Impurities in the semiconductor crystal determine whether the material conducts electricity primarily by the movement of excess electrons (n-type) or, conversely, by the shuffling of electrons through vacant spots, called "holes," where a deficiency of electrons leaves an effective positive charge (p-type). The operation of all semiconductor devices depends on the reaction of electrons and holes to electromagnetic fields applied at interfaces of n- and p-type regions. At the interface, when electrons migrate into the p-type region, where conduction is primarily by holes, they are said to be minority carriers. The same term is used for holes in an n-type region, where electrons carry most of the current. Shockley's junction transistor featured alternate layers of semiconductors that are intrinsically n- and p-type sandwiched together. Bardeen and Brattain's transistor, called the "point-contact" model, used the electric activity of the metal leads that connected the crystal to the circuit to induce only small regions of n- or p-type conductivity in a single semiconductor of one intrinsic type. The junction transistor boasted a larger interface between n- and p-type semiconducting crystals.

After thinking and experimenting for over a year, Shockley published his theory of the junction transistor in June 1949.[34] His paper made it clear that the action of the junction transistor, unlike the point contact type, depended heavily on the flow of minority carriers. Shockley saw this confirmed experimentally in an "existence proof" junction transistor made for Shockley by Bell chemists Morgan Sparks and R. M. Mikulyak in April 1949. The lifetime of minority carriers was quite limited, however, in this trial device made of polycrystalline germanium. By contrast, measurements showed that in Teal and Little's single crystal germanium, minority-carrier lifetimes exceeded that of their polycrystal counterparts by 20 to 100 times.[35] This characteristic gave single crystals an obvious advantage over polycrystals in achieving Shockley's design. Teal and Sparks began collaborating, and for the first time, Teal's research came out from underground. Once Bell's perspective on single crystals matched Teal's own, the crystals' applicability to the junction transistor was recognized. Bell dedicated a laboratory for crystal growing at the end of 1949, and Teal's communication with the transistor development team increased.[36]

Working with Morgan Sparks, Teal began his first work explicitly on the transistor, trying to produce the crystals needed for Shockley's proposed junction device. To create crystals that had alternating n- and p-type layers, as demanded by Shockley's design, Teal modified his crystal-pulling machine to accept impurity pellets into the melt while the single crystal was growing. This enabled him to "dope" the germanium crystal, that is, add impurities to change the germanium crystal's conductivity from n-type to p-type and then back to n-type, in precisely controlled stages. With this improvement, Shockley, Sparks, and Teal were able to grow the first n-p-n junction transistor on 20 April 1950.[37] One year later, Shockley, Sparks, and Teal achieved a useful and reliable junction transistor.[38]

Shockley and Teal disagree when recalling the intensity of the transistor development effort during this period. Shockley's claim that "The efforts to improve junction transistors were practically negligible at the Laboratories" is opposed by Teal's impression that "At no time did the enthusiasm for the double doping program lag."[39] The disparity between the two scientists' perceptions most likely stems from the difference in the roles they played in the development process. Shockley, the theoretical physicist, might not have understood the formidable technical challenges Teal faced in refining his crystal pulling procedure. The deceptively simple concept underlying Teal's technique hides a great number of practical complications that made crystal pulling a difficult enterprise: Precise temperature control was critical; to achieve the necessary crystal uniformity, Teal had to determine how much agitation of the melt and vibration of the seed were required; nonuniformities in the distribution of the impurity in the melt and the tendency of the crystal to change resistivity as it grew added further considerations; and so on. Only through painstaking experimentation was Teal able to successfully pull crystals suitable for junction transistors.[40] (See Figure 3.)

Figure 3.   Teal at his crystal-pulling apparatus.

Back to Texas

Manufacture of transistors as circuit components began slowly in the
1950s. Of the eight companies making transistors in 1952, the largest was
AT&T's own Western Electric, which began making transistors of the point
contact type at its Allentown plant the year before.[41] Western Electric
produced most of the 90,000 transistors made in 1952, but all of these were
for internal use by AT&T. AT&T made no attempt to expand into other
electronic enterprises by marketing transistors or selling transistorized
products. Efforts along those lines would not only have violated a long-
standing company tradition to stay out of such markets, but might also
have drawn unwanted attention to the company. AT&T had good reason
to seek a low profile with regard to new solid state technology. In 1949 the
US Government began antitrust proceedings against the telephone giant.
AT&T, which was eager to maintain its monopoly and integrated structure,
may have been willing to sacrifice fully exploiting the transistor's potential
to avoid hurting its position either in a court of law or in the court of public
opinion.[42] The company decided its best strategy was to collect royalties
from electronics companies that it would license to use the transistor
patent. In the fall of 1951, AT&T offered nonexclusive licenses for a
nominal fee, $25,000 advance against future royalties.

An early respondent to AT&T's offer was Geophysical Service Incorporated (GSI), a Texas-based manufacturer of seismic equipment. This firm,
which had almost no experience in manufacturing electronics, took an
interest in transistors as the cornerstone of a comprehensive program to
redefine itself for the post-war era. GSI had specialized in oil-exploration
services and equipment since its founding in 1930. It developed a strong
reputation within its field, making contracts with the major oil companies
and eventually expanding into the oil business itself. This growth pattern
was altered, however, by World War II. The company's technologies for
remote sensing and location were useful to the military for applications
such as finding land mines. GSI became a supplier of military equipment,
selling $1.1 million of equipment to the armed forces during the war years.[43]
The turn to military contracting not only expanded GSI sales, but also
appealed to the business sense of its principal management. Company
President Erik Jonsson and General Manager Pat Haggerty predicted that
postwar geopolitics would compel the US to maintain a high level of
military procurement. They decided to capitalize on this by expanding
GSI's regular enterprise to include military hardware.

Jonsson and Haggerty's foresight rewarded the company well. GSI
vigorously courted the military market, adding to its product line such
electronic technologies as submarine detection equipment, radar-con-
trolled bomb sights, and airborne and ground radar. Spurred by the Korean
War, sales for GSI were over $15 million in 1951, up from $2.3 million in

1946. Most of the income resulted from military contracts.[44] With this success pointing the way for the future direction of the company, Haggerty and Jonsson began the 1950s looking for new opportunities for GSI.

Bell Lab's development of the transistor provided a special opportunity for GSI. The transistor, which promised great miniaturization and ruggedness, seemed ideal for meeting the military's desiderata for the hardware it purchased.[45] Haggerty's vision, however, called for GSI to do more than simply employ transistors. He believed that in order for his company to become an important electronics firm, it would have to be involved at the component level and not merely assemble systems. Jonsson's desire that GSI start large-scale manufacturing primed Haggerty's interest in obtaining a license to produce the new solid state amplifiers invented at Bell. In 1951 Haggerty had one of his engineers, Bob Olson, test several sample transistors. Olson's investigations convinced Haggerty to pursue transistor manufacture. He met with AT&T representatives and, by the fall of 1951, submitted his $25,000 advance for the right to get into the transistor business. Haggerty, Jonsson, and the other GSI executives reorganized their company to better suit its new interests. They lowered GSI, with its tradition of seismic exploration equipment, to subsidiary status, and created above it, on 1 January 1952, a new parent company named Texas Instruments.

Texas Instruments (TI) faced significant obstacles in its route to become a large company based on transistor manufacture. In 1952 the transistor still ran a poor second to vacuum tubes in almost every measure of performance.[46] Transistors could not approach the frequency response or power output of tubes. The solid state device was more expensive than tubes and transistors were only slightly smaller than the subminiature-type tubes that were readily available. In some cases, tubes held the edge over transistors in size also. In 1955 General Electric made a dramatic show of the small size of its ceramic tubes by placing one inside the casing of a standard transistor.[47] It was clear that a major R&D effort awaited any firm that planned to make money in semiconductors. Haggerty planned a two-pronged approach to overcome these problems: research and development to improve the transistor's characteristics and a high-visibility showcase product to promote the transistor's acceptance.

Texas Instruments began by attending a landmark symposium organized by AT&T in the spring of 1952. The symposium provided a complete disclosure of AT&T's experience with transistor theory, design, and manufacture. For eight days representatives from each of AT&T's transistor licensees, a total of forty firms (twenty-six of them American) heard lectures on solid state physics and toured the Murray Hill and Allentown facilities.[48] Each participant received a detailed exposition of the information covered in the seminar in a multivolume set of books entitled *Transistor Technology*. The symposium was intended "to enable qualified engineers to set up equipment, procedures, and methods for the manufacture of these [transistor] products," but Haggerty knew that he had to do better than this.[49] Certain that transistor technology could not, at its present state, dent the tube business, Haggerty was looking for inroads to improving the state-of-the-art.

Haggerty found the opportunity he sought when he met with Teal at the New Jersey symposium. Teal, who had authored one of the chapters in *Transistor Technology* on pulling crystals, made a presentation on his topic to the symposium attendees. Haggerty was impressed with Teal, and later met with him to offer him the chance to return to his native Texas and found his own research laboratory at TI. The offer appealed to Teal for both professional and personal reasons. Professionally, it presented Teal with the opportunity not only to do research, but also to manage it. The difficulties he had experienced at Bell must have made this seem an attractive prospect. Fondness for his native state and an illness in his family also favored a return to Texas. At the end of 1952, he relocated in Dallas and began to build a research laboratory.

Teal's responsibility, phrased by Haggerty in the corporate language of "goals" and "ambitions," included such imperatives as the following: pioneer the development of technology; integrate diverse scientific disciplines and advanced technologies to create projects of great potential impact on TI's future; create new ideas and concepts through bold research.[50] In practical terms, that meant pulling together a diverse and highly competent staff of research scientists and providing them with sure direction. Teal visited other companies and universities with strong programs to entice talented researchers to follow him back to the small Texas company. Teal's transistor work had bestowed him with a reputation that aided his recruitment efforts. He was able to attract top talent to his fledgling lab, including a chemist from Standard Oil, Willis Adcock, and a recent Oxford graduate in physics, J. Ross Macdonald.[51]

Transistor research was a wide-open field in 1953. The outstanding problems confronting investigators included the chemical composition of the semiconductor material to be used, the purity of semiconducting crystals, and the method for fabricating the transistor. There were at least as many different ideas on how to proceed as there were laboratories to pursue them. At Bell Labs important progress had been made in purifying germanium with the zone-refinement process developed by W. G. Pfann in 1951. The following year J. E. Saby at General Electric developed an entirely new way to prepare junction transistors that allowed one to eschew the crystal-pulling apparatus entirely. Saby composed alternate n and p layers by alloying dots of indium to a base of germanium. This technique, which offered greater frequency range and current handling (provided the germanium base was thin), was adopted by RCA and Raytheon. Philco, a Philadelphia electronics firm, mastered a jet-etching technique for machining exceptionally thin germanium bases in 1953, giving their alloy transistors the highest frequency response yet.[52] These companies avidly researched the alloy technique because they believed it held promise for making transistors that were not only superior but also cheaper—the pulling process for the grown-junction transistor was notoriously slow.

At Texas Instruments, however, Teal's presence led the company toward the crystal-pulling technique that he had pioneered, a fact that proved decisive in moving that firm ahead of the other transistor licensees. Under the direction of Mark Shepherd, chief engineer for semiconductor design, TI began pulling crystals immediately after the Bell symposium. Working on a machine dubbed "Old Betsy," Shepherd's group made their first transistor in June. By the end of the year they had found a customer for their product: The Gruen watch company purchased ten of these point-contact transistors. In 1953, production increased to 7,500, spurred by an order from the Sonotone hearing aid company. These transistors were expensive, costing between $10 and $16 each, but this did not concern Haggerty excessively. At such small production volumes, competitive edge from superior technique was an almost meaningless concept. The critical issue for Haggerty was generating a demand for transistors. He cared little about optimizing the performance of transistors as laboratory devices; he wanted to create the impression that transistors were a proven technology and simultaneously establish Texas Instruments as the preeminent company to fill the ensuing demand. It was vital, then, that Texas Instruments act quickly.

In the spring of 1954, Haggerty set his engineers on a breakneck development project to design a transistor radio that would fit in a shirt pocket and cost no more than $50. After extraordinary effort Haggerty's engineers met his target of a Christmas release date. The radio, the TR-1, manufactured by the Regency division of the IDEA Company of Indiana, sold 100,000 units in its first year. At four transistors per radio, the phenomenal sales of the TR-1 achieved the market breakthrough for transistors that Haggerty sought.[53]

The TR-1 played an important role in fixing the transistor in the mind of the consumer and the commercial manufacturer, but Texas Instruments had been, from the beginning, more interested in the military market. Here, too, the company's commitment to Teal's crystal-pulling technique for growing junction transistors gave it a crucial advantage over its competitors. Certain military applications, the sort of high-performance weaponry emphasized by Eisenhower's "New Look" at military strategy, demanded transistors that functioned reliably at high temperatures.[54] This requirement pointed to silicon, with its high melting point. Eager to include the military on their list of customers, most semiconductor manufacturers began working at developing a silicon transistor.[55] Most research laboratories approached the problem using the alloy method of fabrication, an unfortunate choice since alloyed silicon transistors proved exceptionally difficult to develop. However, by working with the grown junction technique, Texas Instruments was able to market a working silicon transistor well in advance of any other company, effectively monopolizing this large military market for four years.

Possessing a familiarity with silicon from wartime work, Teal had been interested in silicon crystals since his early days working with Little at Bell Labs. In experiments conducted in the late 1940s, Teal found silicon less manageable than germanium. The higher temperature necessary to melt silicon made it prone to react with atoms in its environment, spoiling the perfection of the crystal. Patient work with the element paid off slowly. In 1952 Teal, still at Bell, worked with Ernie Buehler to grow silicon junctions.[56] Functioning transistors had remained elusive, however. By 1953 most research on the silicon transistor involved alloying techniques. Others sought instead to "leapfrog" silicon entirely and make high-temperature transistors out of intermetallic compounds, such as gallium-arsenide. Teal rejected both of these approaches as introducing additional complex practical problems. He placed Willis Adcock in charge of developing a grown junction silicon transistor. By the spring of 1954, Adcock, together with Morton Jones, J. W. Thornhill, E. D. Jackson, Ray Sangster, and Boyd Cornelison, solved the purification problems and mastered the pulling technique.

Teal announced his team's success at the Institute of Radio Engineers national conference on 10 May 1954 with a paper titled "Some recent developments in silicon and germanium materials and devices." The session in which Teal gave the paper was marked by a succession of grim-faced scientists solemnly confessing that the silicon transistor was still years away. Indulging a flair for drama encouraged by Haggerty, Teal ended his modestly titled paper by announcing that "contrary to the opinions expressed in this morning's session, this [the production of the silicon transistor] will begin immediately" and then reached into his pocket exclaiming, "I happen to have some here." (See Figure 4.) The excitement that ensued among the audience foreshadowed the heady growth that Texas Instruments enjoyed over the rest of the decade. Texas Instruments saw sales rise overall from $27 million in 1953 to $233 million in 1960. This environment of expansion was ideal for a director of research.

As Haggerty predicted, military sales were valuable for Texas Instruments, but they were not the sole cause of the company's spectacular growth. Commercial sales of germanium transistors were significant. Texas Instruments entered into an agreement with IBM in 1957 whereby they sold 41,000 transistors for each model 7070 computer. Sales of semiconductor rectifiers also were profitable. Targeting all semiconductor markets, Texas Instruments diversified its product line by maintaining a steady pace of incremental improvements in products, processes, and testing equipment. In 1958, for example, the company announced over 100 new semiconductor products—57 of them meeting military specifications—which improved on past offerings with a wider range of operating parameters and enhanced reliability. Funding for research, development, and engineering that year equaled $16.25 million, half of which was derived from contracts, principally from the federal government.[57]

Figure 4.   Two of Texas Instruments' first silicon transistors.

Texas Instruments' marketing triumphs of the 1950s reflect a corresponding string of research successes in Teal's Central Research Lab (CRL). In addition to transistors, TI manufactured silicon components such as a cell to convert solar energy to electricity (introduced in 1955) and the sensitor, a silicon resistor, made in 1957. Refinement of the procedures that CRL developed to attain ultrahigh-purity silicon enabled TI to market the material in bulk form to other electronics firms, as TI joined Du Pont in 1957 as the only suppliers of this high-grade substance. The need for exact control in this chemical process stimulated development at CRL of high-precision test and measurement equipment. The lab invested heavily in research with intermetallic compounds such as gallium-arsenide and indium-antimonide, materials that found use in diodes, transistors, and infrared detectors. Other CRL projects included work on fuel cells and the automation of electric power systems.

Teal's success at running the Central Research Laboratory emboldened him to seek a larger role in the management of Texas Instruments. According to Howard Sorrows, CRL's director of new product development, Teal contributed significantly during the 1950s to the development of OST (Operations, Strategies, and Tactics), a rigorous company-wide management system that Texas Instuments instituted in the early 1960s. OST required certain preliminary planning operations before work on a new project began, such as identification of the clients for a proposed product, drafting of a detailed plan of development, and securing written agreements between the internal divisions that planned to participate in the development effort. Sorrows recalls that Haggerty derived these aspects of OST, along with other features such as zero-based budgeting, from the system that Teal implemented at CRL.[58]

Teal downplays his management efforts at Texas Instruments, perhaps because they generated some tension among the company's principals. Haggerty and Shepherd received Teal's management recommendations only coolly. They preferred that he concentrate his labor on scientific matters and take a more passive role in general management. The pair regarded those qualities that they respected in Teal as a scientist, namely, his analytic abilities and his thorough methodology, as liabilities in corporate management. Haggerty had concluded early on that executive responsibility often demanded that managers make decisions based on only incomplete information and imperfect comprehension. He worried that Teal was not inclined to operate in this mode.[59]

An organizational dispute over the development of the integrated circuit exemplifies this conflict over Teal's proper role within TI. The integrated circuit, Texas Instruments' most important product since the silicon transistor, emerged not from Teal's CRL but from another unit within TI called the Semiconductor Products Division (SPD). This unit, which was organized in 1954 under Shepherd to oversee the refinement of semiconductor products for the marketplace, was the development counterpart to Teal's research laboratory. Although the two divisions worked closely, they were operationally quite distinct.[60] When Haggerty and Shepherd assigned the responsibility for the integrated circuit (IC) to the SPD, they shut out Teal's unit from the project.

The decision dismayed Teal. He felt that the IC project belonged in the CRL. He traced its origin back to research that Richard Stewart performed at CRL in 1956. Stewart, working under Adcock, created arrays of diodes in single wafers of silicon for use in microwave detectors. In 1957 Adcock transferred to the SPD to work on microminiaturization and shortly thereafter hired Jack Kilby. Kilby, in 1958, built the first integrated circuit, fabricating multiple electrical components, such as resistors, capacitors, and conductive connectors, out of a single crystal of silicon.[61] Although Adcock has emphasized the differences between Stewart's work and Kilby's—the key one being that Stewart never attempted to form different electrical components out of the single crystal—Teal felt that integrated circuit research began at the CRL.[62] Moreover, he believed that the project would benefit from a sounder scientific understanding. Haggerty and Shepherd, however, were eager to push development of the IC. They overrode Teal's desire to put CRL staff on the project.

These developments frustrated Teal. It discomforted him to be positioned on the pure science side of the basic-versus-applied research debate that is so frequently an issue for high-tech firms. Teal fully comprehended Haggerty and Shepherd's desire to market the integrated circuit as soon as possible (their haste notwithstanding, Texas Instruments' innovation collided in patent court with similar work done by Robert Noyce at Fairchild Camera's electronics department). Teal was a pragmatic individual. Indeed, his own inclination, the spur that drove his interest in germanium, was to find uses for things. It surprised Teal, then, to find that, despite his own inclination toward pragmatism, he was seen by Texas Instruments as being, as Shepherd expressed it, "sort of Bell Labs-like. A little on the academic side."[63] If at Bell, Teal was held back for a time by the scientific myopia of his superiors, at Texas Instruments he could chafe against the explicit business perspective that underlay crucial organizational decisions. He took the event as a lesson; his writings and activities of the late 1950s reflect a heightened attention to corporate interests.

Teal had to undergo several educations to learn the subtleties of his position as a research manager in a commercial enterprise. He was acutely aware of the differences in his responsibilities at Texas Instruments as compared with Bell and undertook to equip himself for his new duties as best as possible. He remembers, "When I got to be a boss instead of a scientist working mostly by himself . . . I began communicating, being interested to hear what everybody else was doing. . . . So that got me involved with working in the Dallas-area scientific organizations."[64] Teal joined the American Management Association, Research Division in 1953. That same year, he joined the Dallas chapter of the Institute of Radio Engineers (IRE). He quickly took a leadership role in that organization and in 1956 was named chairman. As a local leader for the IRE, Teal organized regular meetings, started a local magazine called *Direction,* and planned social occasions that gave him opportunities to meet and trade concerns with the growing community of Texas electrical engineers.[65]

Teal had belonged to the American Chemical Society since 1927; his involvement with the IRE was an acknowledgment of his new relationship to electronics. His close association with Haggerty reinforced his desire to become active in the IRE. (See Figure 5.) Haggerty was named president of the Institute in 1962, the year preceding its merger with the American Institute of Electrical Engineers to form the present-day Institute of Electrical and Electronics Engineers. Teal favored the merger, expecting that the consolidation would expand the interests of members of each society. He participated by serving on a two-man study committee on the merging of the sections, considering the finances, geographical boundaries, and sectional publications.

In addition to these activities, Teal was enlisted as a member of the editorial board for the IRE, a national position that he held between 1958 and 1961. He also joined SWIRECO, the IRE convention organization for

Figure 5. Patrick E. Haggerty (1914–1980), President of Texas Instruments and President of the IRE for 1962.

the Southwest, in 1955. Four years later, he was vice chairman. That same year, Teal was appointed director-at-large of the IRE, a position he reclaimed by election in 1962.

Teal found that his professional society affiliations presented him with an opportunity to express his opinion on issues that were becoming more important to him. One of these was education. At the 1958 IRE National Convention, he presented a paper entitled "Broadening the horizons of the engineer." This lecture stressed not only the changes in physics and chemistry that were transforming the world of the engineer, but also the general political and economic forces in the world scene, the growing importance of industrial research, changing technology and corporate needs, and societal needs. He elaborated on the value of education for American industrial growth and stability.[66]

The present dilemma in education indicates clearly the need for the engineer to be effective not only on a specific technical job but also in society. One of the major needs of our times is to make science an integral part of our culture just as it is an integral part of present day technology. Failure of our educational system to achieve this no doubt accounts largely for the unfortunate fact that the public as a whole has little understanding of science. This offers the engineer a unique opportunity. Since his work is better understood by the public than that of the pure scientist, he is ideally suited to interpret society's dependence on pure science as well as applied science. In doing this the engineer will be rendering a service to his community and nation. This could lead the average citizen to have a clearer understanding of decisions of state involving complex scientific knowledge. The engineer himself, however, to have the interest in such activity, must be educated broadly.[67]

The zeal in America to prioritize education, particularly in science, that followed the Soviet Union's successful launch of Sputnik focused some of Teal's ideas on education. In 1958 he spearheaded the creation of a body called the Council of Scientific and Engineering Societies and served as the chairman of the executive committee of its board of directors in 1961 to 1962. The Council provided service by assisting teachers in extending their education, advocating higher teacher qualifications and salaries, sponsoring scientific lectures for the general public, and promoting student interest in science through activities such as science fairs and science-oriented TV programs. Writing of the Council's purpose, Teal asserted "the value of learning is an end in itself, worthy of great effort," but conceded also that "national security and material wealth as objectives must be of concern to all of us. These are increasingly dependent on activities of the intellect."[68] This social motivation, in contrast to a purely intellectual one, is apparent in activities such as the Council's cosponsorship with the Council on World Affairs of a civil defense symposium, held on 7 and 8 December 1961, at which Edward Teller, the physicist known for his strong political views, spoke.

Teal on Assignment

The concern for large social issues demonstrated by these activities underscored the variety of ways in which Teal could be valuable to Texas Instruments. In the 1960s Haggerty had an opportunity to use some of these extra-scientific talents. The company's explosive growth during the 1950s came to an abrupt halt in 1961. The Kennedy administration, fresh in office, ordered a slowdown of military procurement while it reassessed US strategy. This decrease in the demand for semiconductors exacerbated the problem of overproduction (20 percent in 1961) and led to a brutal price war. For the first time since the appearance of the silicon transistor, Texas Instruments profits were off from the previous year, $9.4 million compared with $15.5 million. The company did not panic—they were still the top supplier of semiconductors and they still made a profit. But the semiconductor recession of 1961 alerted Texas Instruments to the fact that as a large, mature company, it faced an entirely different set of problems that innovation through scientific research could not solve alone.

One of these problems was its overseas operations. Texas Instruments began production outside the US in 1957, when it built a plant in Bedford, England, to manufacture semiconductors. Texas Instruments, Ltd., the TI subsidiary operating the Bedford plant, was a success. By the end of 1960, Texas Instruments made 12 percent of its sales outside the US and operated seven manufacturing plants in England, France, Italy, Germany, and Holland.

With so much activity outside the United States, Haggerty began to wonder if the Dallas Central Research Laboratory was adequate for all of the firm's interests. Thinking that better ties might be established with European scientific talent if Texas Instruments operated a research lab on that continent, Haggerty asked Teal to investigate whether a branch lab should be built, and if so, where. In 1963 Teal, named International Division technical director, left with his family for an extended tour in Europe. He worked in London, Paris, and Rome, visiting government laboratories, industries, and universities to discuss research with officials, managers, scientists, and professors. Teal ultimately decided that the logistical complications of managing research in two sites so distant from one another outweighed any possible advantages a European research lab would offer. Even with the negative decision, the trip had great value for Teal (apart from the cultural exposure that he thirstily absorbed). Teal shined in the opportunity to display his sensitivity to the complex interactions between science, technology, and industry.

Teal's sophisticated grasp of these issues was not lost on Haggerty, or on certain other keen observers. Late in 1964, when Teal was still in Europe, the director of the National Bureau of Standards, Allen Astin, and his deputy director, Irl Schoonover, approached Haggerty, asking if TI could "loan" Teal to the Bureau for a two-year stint as director of a new Bureau division called the Institute for Materials Research (IMR). The Institute was to be "the principle focal point in the federal government for assuring maximum application of materials sciences to the advancement of technology in industry and commerce."[69] Its head was to be responsible for the stimulation of industrial capacity through rationalization of the production and distribution of scientific data about materials. Bureau officials sought an astute materials scientist with experience at organizing a research laboratory and an understanding of the needs of the industrial community.

Astin and Schoonover recognized Teal as an ideal candidate. Haggerty, agreeing with their judgement, telephoned Teal in Rome and put the request to him. Teal quickly assented, excited by the challenge of organizing a new research institution. In January 1965 he went to Gaithersburg, Maryland, to begin work at the National Bureau of Standards. (See Figure 6.)

Figure 6. The National Bureau of Standards buildings in Gaithersburg, Maryland.

The Bureau had a long tradition of assisting industry through scientific research. It was formed in 1901 under the Department of the Treasury, principally to deliver accurate measurements, and was moved under the auspices of the Department of Commerce when that agency was created in 1903. The change in parent administration encouraged the Bureau to pursue functions beneficial to industry, such as standardization of parts and products or promotion of new and improved materials.[70] The rapid growth of science during World War II and the subsequent Cold War expanded the Bureau's activities even further. This trend continued into the 1960s, but tight budgets in those years prevented the Bureau from growing in proportion to its responsibilities. The Bureau needed to take steps to improve the effectiveness of its operation.[71]

In the drive for a more cost-effective agency, Astin heeded the recommendation of a National Research Council advisory committee to take a new approach to organization.[72] He identified the missions that were shared by the numerous different divisions of the Bureau and regrouped these divisions along the lines of these common objectives. In 1964 all research within the Bureau was divided into four decentralized units: the Institute for Basic Standards, the Institute for Applied Technology, the Central Radio Propagation Laboratory, and the Institute for Materials Research.

Astin formed the last of these, the Institute for Materials Research, by pulling together six distinct divisions within NBS: the laboratories for analytic chemistry, polymers, metallurgy, inorganic materials, reactor radiation, and cryogenics. In each of these six areas, the Institute offered valuable services to its industrial clients and other government agencies. Its program included the development of techniques for the study and manufacture of materials; identification and measurement of meaningful physical and chemical properties of materials; on-site calibration services to scientific and industrial institutions; consultation services on materials; and identification of critical materials problems that obstructed major national goals.

Teal thrived in his position as the first director of the IMR.[73] His initial task was to execute the internal changes that would bind the formally independent divisions of the Institute into a coordinated research organization. Initially Teal did not need to do any recruiting, because the Institute, unlike the Central Research Laboratory at Texas Instruments, had a staff of over 600 when he arrived, 370 of whom were professionally trained, including 160 Ph.D.s. Still, his skills in evaluating scientists' abilities and matching them to the jobs proved critical in launching the Institute.[74]

As leader of the Institute, Teal attacked the problem of how to maximize the value of his laboratory's resources to industry with the same spirit he had shown on more scientific problems earlier in his career. He held a five-year management planning review on 4 April 1966. Proceeding methodically, Teal first outlined national needs and goals, then identified customers and program outputs, assessed the impact of Institute programs on national goals, and pointed out the strengths and weaknesses of the available resources and plans for achieving their objectives. The review was a huge success. Within a year the entire Bureau was conducting similar sessions.[75]

Teal developed his assessment of national research goals through consultation with leading U.S. industrial firms. From these meetings, he decided which materials studies were most crucial by considering impact evaluations prepared by a staff of technical economists that he brought to the Institute. Directed by this input, IMR developed techniques to prepare such materials as hyperpure copper and crystals of other metals, new silicon compounds, superconducting ceramics, and vinyl-like polymers.[76] Other techniques developed there helped scientists determine the physical characteristics of materials. Examples include an assembly to rotate a sample of steel in a neutron environment in order to sense the presence of oxygen in steel, and a method to identify trace elements in substances by measuring the ratio of residual current and thermoelectric power in materials at high and low temperatures. The sort of practical objectives underlying the work of the IMR is revealed by the report of one success in the Bureau's 1965 annual survey of technical highlights: "IMR scientists desired an improved spectroscopic technique for chemically analyzing the high-temperature alloys being used in increasing quantities in modern technology. The technique [which combined the principle of isotope dilution with electrolytic separation and determination by a spark source mass spectrometer] is readily adapted to existing instrumentation and can save industry time and money because the number of primary standards needed to calibrate the analytical equipment is reduced."

One of the Institute's most important contributions was the preparation and distribution of standard reference materials. These were samples of substances, each carefully prepared, that laboratories could use as a reference for precision adjustment of their instruments. The Bureau sold these materials (known as standard samples before Teal's arrival) to academic and industrial laboratories. Ownership of a reference material enabled a laboratory to calibrate its equipment itself, on-site, sparing it the trouble and expense of seeking an outside contractor. By using an NBS reference material, customers were assured that their processes conformed to a national standard, clearly valuable for its reproducibility and uniformity. When Teal surveyed reference materials buyers in 1965 for reactions to a proposed downscaling of the reference materials program, he was inundated with pleas to reconsider. The emphatic responses stressed how much industry appreciated the service and suggested that any shortfall in funding should be met by raising prices.

The references sold briskly. In 1965, the Institute sold over 70,000 samples to more than 2000 different companies.[77] Clients included the metal, petroleum, automotive, cement, chemical, rubber, plastic, pharmaceutical, and transportation industries. Other materials, which set standards for radioactivity, were useful in calibrating civil defense equipment. A study Teal conducted to assess the demand for new types of reference samples indicated that industry, to develop all the products and processes made possible by present technology, urgently needed over 150 new reference materials.[78] With budgets tight, this job was beyond the Institute's capacity. Teal responded by prioritizing critical needs and focusing limited resources where they would benefit most. In 1966 the Institute offered 91 new reference materials, swelling the total available to more than 600.[79] Sales in 1967 exceeded $1.2 million, representing more than 70,000 samples sold to over 8,000 customers.[80]

The Standard Reference Materials were just one method by which the Institute disseminated scientific information. In addition, the Institute published papers and books and also organized conferences and symposia on topics such as ceramics, corrosion, and the growth of crystals. Teal organized an annual series of IMR symposia as a focused effort to transfer knowledge about a particular subject among different scientific groups from around the world. The first symposium, held in 1967, considered the characterization of trace elements in materials.

Teal regarded these measures to offer quality scientific results to laboratories that needed them as simply the first step toward the Institute's goal. He envisioned the Institute as the progenitor of a vast, international, and intercorporate system of coordinated research activity, and he spoke of consolidating all research results into an international bank of scientific data. Citing a cost of $40,000 for every good technical research paper, Teal pointed out that eliminating worldwide research redundancy would relieve industry of an expensive burden that was a drag on productivity. He called on all interests to collaborate in this mutually beneficial mission and went on to suggest that underdeveloped nations could be granted access to the data network, offering them a chance for industrial growth and economic stability.

These suggestions, which he offered to members of the Industrial Research Institute (IRI) at a meeting in Detroit on 11 October 1966, were Teal at his most idealistic. Though he found the task of rationalizing the activity of competing companies and nations along lines of enlightened self-interest too much to accomplish in his two years as head of the IMR, he nevertheless did take a few modest steps toward his brave new world. After his talk to the IRI, representatives of the group came to Maryland to observe the Institute's program. Teal was also able to visit India, Pakistan, and Israel as part of an IMR delegation to instruct those countries in developing standard reference materials and other engineering standards.

. . . We will make faster progress in solving our national problems by recognizing that they are fundamentally different from the many problems solved by science and technology during the last few decades. The essential difference is that such magnificent achievements as penicillin and radar and supersonic flight were problems that could be solved by the biological and physical sciences alone. . . . In contrast, today's problems demand not only technological decisions, but the simultaneous application of many diverse disciplines and nondisciplines; in this regard, the new problems are terribly complex. . . . It seems to me that there are four keys to each of the complex problems that face us—four basic conditions that, taken together, will assure our success: First, we must articulate our national prime objective and measure progress toward it. Second, our approaches to subsidiary goals that serve this prime objective must be both interdisciplinary and extradisciplinary. Third, we must take systems approaches and avoid piecemeal solutions. And fourth, we must make the problems attractive to private industry so that industry will take the lead under government direction.[81]

Teal's tenure at the National Bureau of Standards ended in 1968, half a year after he was scheduled to depart. Having set the IMR off to a strong start, Teal left Maryland and returned to Dallas for his final years with Texas Instruments. He assumed the office of vice president and chief scientist for corporate development, which he held until his retirement in 1972. He enjoyed the responsibility of his position, plotting the technological course of a company he was instrumental in building.

Teal stayed active with the company after his retirement, consulting for another five years. He also consulted occasionally for the National Academy of Science, the National Academy of Engineering, the National Bureau of Standards, the Department of Defense, NASA, and the National Science Foundation between 1972 and 1978.

Over the years Teal has been much honored for his achievements. In 1968 he received the highest award of the IEEE, the Medal of Honor. Teal won the Patent, Trademark and Copyright Research Institute's Inventor of the Year Award for 1966, was elected Member of the National Academy of Engineering in 1969, was awarded the Academy of Achievement's Gold Plate Award in 1967, received honorary degrees from both Baylor (L.L.D., 1969) and Brown (Sc.D., 1969), was named an Omicron Delta Kappa Outstanding Baylor University Alumnus, received the American Chemical Society Award for Creative Invention (1970), and an IEEE Centennial Award (1984), to name just the most prestigious of Teal's citations.[82] It is a long, varied, and distinguished list—matching the career of the man who earned it.

Conclusion

Gordon Teal's contribution to solid state electronics, the monocrystals of germanium and silicon that opened up the field to practical use and commercial viability, guarantees him a high place in the history of technology. His story is interesting also for what it shows about science and engineering.

Historian of technology Edwin Layton has written of a nineteenth century "scientific revolution of technology" in which "technological problems could be treated as scientific ones; traditional [craft] methods and cut-and-try empiricism could be supplemented by powerful tools borrowed from science."[83] He alludes to a convergence of scientific and technological research, one that is commonly taken to be a matter of engineers "upgrading" themselves to the more "scientific" methods of mathematization and controlled experimentation. Teal's career suggests that this convergence was not necessarily as unidirectional as is generally assumed.

Teal started as a scientist. His studies at Brown were straightforward chemistry, and his early work—at Bell Labs and at Columbia with Harold Urey—bore the same unmistakable scent. He was, however, possessed of certain values and characteristics—a pragmatic enthusiasm for utility, an appreciation for the physicality of objects that sat above the desire to understand them theoretically, and a commitment to invent techniques to create perfect specimens—which fit the profile of an engineer. Guided in his work by an irrepressible inventiveness, superlative experimental skill, and an educated instinct, Teal was a scientist who thought and functioned as an engineer. He represents one important example where scientific practice absorbed the advantages of engineering practice.

Teal's style was particularly striking in the context of solid state research. Most interpretations of the development of the transistor emphasize the significance of theoretical advances to this science-based technology. Theory was a potent tool, but at the same time Teal's practice of employing dexterous technique in patient experimentation guided by extra-theoretical instincts was not irrelevant. Even as theory-rich physical chemistry was increasing in importance to the electronics industry, the work of one leader in the movement echoed the engineering traditions of an earlier age.

In his later career Teal continued to work as an engineer. His management of the research enterprises at Texas Instruments and the National Bureau of Standards offered him the chance to design and build in a way that was more abstract, but no less real, than he had experienced as a scientist. Then, as before, he sought to collect and perfect. Making the most of the resources available to him, Teal was always looking for the right material.

---

[1] Robert R. Williams, "Chemistry in the telephone industry," *The Bell System Technical Journal*, vol. 9, 1930, p. 603.

[2] Richard L. Petritz, "Contributions of materials technology to semiconductor devices," *Proceedings of the IRE*, vol. 50, 1962, pp. 1025–1038.

[3] Most of the information about Teal's background comes from an extended oral history interview with Teal conducted by the author 17 to 20 December 1991. This interview is the basis for much of the content of this article. It is supplemented, as noted in the footnotes, by various published and unpublished materials. An edited transcript of the interview is deposited in the archives of the Center for the History of Electrical Engineering.

[4] See, for example, Michael Eckert and Helmut Schubert, *Crystals, Electronics, Transistors*, translated by Thomas Hughes (New York: American Institute of Physics, 1986), p. 65.

[5] Interview 1991, p. 29.

[6] See G. K. Teal and C. A. Kraus, "Compounds of germanium and hydrogen. III. Monoalkylgermanes. IV. Potassium germanyl. V. Electrolysis of sodium germanyl," *Journal of the American Chemical Society*, vol. 72, 1950, pp. 4706–4709.

[7] From "Patterns of creative action in a research career," a speech given at a meeting of the American Academy of Achievement in Dallas, Texas, 15–17 June 1967, to 350 high school students from 48 states.

[8] Harold C. Urey and Gordon Teal, "The hydrogen isotope of atomic weight two," *Reviews of Modern Physics*, vol. 7, 1935, pp. 34–94.

[9] Albert Abramson, *The History of Television, 1880 to 1941,* (Jefferson NC: McFarland & Company, 1987), p. 101.

[10] R. W. Burns, "The contributions of the Bell Telephone Laboratories to the

early development of television," *History of Technology*, vol. 13, 1991, pp. 181–213.

[11] V. Zworykin et al., "Theory and performance of the iconoscope," *Proceedings of the IRE*, vol. 25, 1937, p. 1091.

[12] Teal lab notebook 13106, AT&T Archives, Warren NJ. See also patent numbers 2,641,533; 2,598,317; 2,681,886; 2,650,191; 2,682,501; 2,662,852; 2,598,318; 2,596,617; and 2,662,274.

[13] Patent number 2,650,191.

[14] Patent number 2,641,553.

[15] Harley Iams and Albert Rose, "Television pickup tubes with cathode-ray beam scanning," *Proceedings of the IRE*, vol. 25, 1937, pp. 1066–1067.

[16] Michael Wolff, "The R&D 'Bootleggers': Inventing against the odds," *IEEE Spectrum*, vol. 13, no. 7, 1975, p. 39.

[17] Teal's lab notebook 15202, AT&T Archives, Warren NJ.

[18] S. Millman, ed., *A History of Engineering and Science in the Bell System: Communications Sciences (1925–1980)* (AT&T Bell Labs, 1984), p. 142.

[19] The figures for Bell are from Burns, "The contributions of the Bell Telephone Laboratories . . . ," p. 209. The RCA figure is quoted from David Sarnoff in Andrew F. Inglis, *Behind the Tube: A History of Broadcasting Technology and Business* (Boston: Focal Press, 1990), p. 171.

[20] Interview 1991, p. 50.

[21] Teal's lab notebook 15202, AT&T Archives, Warren NJ. See also patent number 2,408,108.

[22] Millman, *A History of Engineering and Science*, p. 145.

[23] Teal's lab notebook 18414.

[24] J. H. Scaff and R. S. Ohl, "Silicon crystal rectifiers," *Bell System Technical Journal*, vol. 26, 1947, p. 28.

[25] Teal, "Single crystals of germanium and silicon—basic to the transistor and integrated circuit," *IEEE Transactions on Electron Devices*, vol. ED-23, 1976, p. 621.

[26] G. K. Teal, M. D. Rigterink, and C. J. Frosch, "Attenuator materials, attenuators, and terminations for microwaves," *AIEE Transactions*, vol. 67, 1948, pp. 419–428.

[27] For an exposition of Bell's results, see Frank R. Stansel, "The characteristics and some applications of varistors," *Proceedings of the IRE*, vol. 39, 1951, pp. 342–359.

[28] For a more detailed account of the story of the transistor's development, see especially Lillian Hoddeson, "The discovery of the point-contact transistor," *Historical Studies in the Physical Sciences*, vol. 12, 1981, pp. 42–76; S. Millman, ed., *A History of Engineering and Science in the Bell System: Physical Sciences (1925–1980)* (Murray Hill NJ: Bell Laboratories, 1983), pp. 71–107; William Shockley, "The path to the conception of the junction transistor," *IEEE Transactions on Electron Devices*, vol. ED-23, 1976, pp. 597–620; F. M. Smits, ed., *A History of Engineering and Science in the Bell System: Electronics Technology (1925–1975)* (Murray Hill NJ: Bell Laboratories, 1985), pp. 1–100; and Charles Weiner, "How the transistor emerged," *IEEE Spectrum*, vol. 11, no. 1, 1973, pp. 24–33.

[29] Teal, "Single crystals of germanium and silicon . . . ," p. 622.

[30] Ibid., p. 623.

[31] William Shockley, "How we invented the transistor," *New Scientist*, vol. 56, 1972, p. 690.

[32] "The creative individual in modern society," presented by Teal on 2 December 1960 as Presidential Address to the General Assembly of the Texas Academy of Science at its annual meeting at Texas Christian University in Ft. Worth, Texas.

[33] Michael Wolff, "The R&D 'Bootleggers' . . . ."

[34] William Shockley, "The theory of p-n junctions in semiconductors and p-n junction transistors," *Bell Systems Technical Journal*, vol. 28, 1949, pp. 435–489.

[35] Gordon K. Teal and John Little, "Growth of germanium single crystals," *Physical Review*, vol. 78, 1950, p. 647.

[36] Teal, "Single crystals of germanium and silicon . . . ," p. 628.

[37] Shockley, "The path to the conception of the junction transistor," p. 616.

[38] William Shockley, Morgan Sparks, and Gordon Teal, "p-n junction transistors," *Physical Review*, vol. 83, 1951, pp. 151–162.

[39] The two quotations are from, respectively, Shockley, "The path to the conception of the junction transistor," p. 616, and Teal, "Single crystals of germanium and silicon . . . ," p. 630.

[40] See chapters 4, 7, and 8 of H. E. Bridgers et al., eds., *Transistor Technology* (Princeton NJ: D. Van Nostrand, 1958) for details on the difficulties of pulling crystals.

[41] Ernest Braun and Stuart Macdonald, *Revolution in Miniature* (Cambridge UK: Cambridge University Press, 1978), p. 59.

[42] I am indebted to AT&T historian Sheldon Hochheiser for this insight into AT&T's business strategy.

[43] "Biography of Pat Haggerty," TI Archives, p. 11.

[44] "Biography of Pat Haggerty," pp. 12–17.

[45] See Thomas J. Misa, "Military needs, commercial realities, and the development of the transistor, 1948–1958," in Merritt Roe Smith, ed., *Military Enterprise and Technological Change* (Cambridge MA: MIT Press, 1985), pp. 253–287.

[46] Ernest Braun and Stuart Macdonald, *Revolution in Miniature* (Cambridge UK: Cambridge University Press, 1978), p. 54.

[47] George Wise, "Research and results: A history of R&D at General Electric, 1946–1978," unpublished manuscript, 1980, pp. 12–13.

[48] Smits, *A History of Engineering and Science . . .* , p. 29.

[49] The quotation is from Smits, p. 29.

[50] "Technical highlights of TI, 1952–1964," a speech presented by Teal to the fall TI Technical Seminar, Semiconductor-Components Auditorium, 12 November 1962.

[51] Mark Shepherd, then TI's Chief Engineer for Semiconductor Design, discusses the importance of Teal's personal reputation in recruiting in an interview conducted with the author by phone on 3 November 1992 (inter-

view held at the IEEE Center for the History of Electrical Engineering archives).

[52] See chapter 6 of Braun and Macdonald, *Revolution in Miniature* for an extended treatment of these developments.

[53] The story of the TI's transistor radio is well told in Michael F. Wolff, "The secret six-month project," *IEEE Spectrum*, vol. 23, no. 12, 1985, pp. 64–69.

[54] See Misa, "Military needs . . . ," and P. R. Morris, *A History of the World Semiconductor Industry* (London: Peter Peregrinus Ltd., 1990).

[55] For example, Philco announced a silicon transistor in January 1954 but also declared that the commercial production was hampered by problems in obtaining enough pure silicon crystals (*Wall Street Journal*, Monday, 10 May 1954). RCA's efforts are described in Herbert Nelson, "A silicon n-p-n junction transistor by the alloy process," presented at the Semiconductor Research Conference of the IRE, June 1954, Minneapolis MN.

[56] G. K. Teal and E. Buehler, "Growth of silicon single crystals and of single crystal p-n junctions," *Physical Review*, vol. 87, 1952, p. 190.

[57] Texas Instruments Annual Report, 1958, pp. 3–5.

[58] Personal communication with Harold Sorrows, 2 February 1993.

[59] Personal communication with Harold Sorrows, 2 February 1993.

[60] The close cooperation of the two divisions reflects the permeable boundary between basic and applied research. When the SPD launched a series of reliability studies they found the precision they required was beyond their capabilities, and so they prevailed on Teal's staff to help them with the sophisticated test instruments that he had in his laboratory. In the other direction, the Semiconductor Products Division provided important assistance for the CRL's own research. For example, in 1954 Boyd Cornelison of the Semiconductor Products Division designed and built an improved crystal-puller for Adcock and his team to study silicon transistors.

[61] See Michael Wolff, "The genesis of the integrated circuit," *IEEE Spectrum*, vol. 14, no. 8, 1976, pp. 45–53.

[62] Adcock's view was presented to the author during phone conversation on 2 February 1993.

[63] Shepherd interview 1992.

[64] Interview 1991, pp. 152–153.

[65] Interview 1991, pp. 160–165.

[66] Teal, "Broadening the horizons of the engineer," *1958 IRE National Convention Record*, Part 10, 24–27 March 1958, pp. 42–45.

[67] Ibid.

[68] Teal, "Report of the Chairman of the Executive Committee of the Board of Directors," 5 June 1962.

[69] Annual Report, Institute for Materials Research, 1965, p. 79. These same words were used by the Department of Commerce to describe their hope for the whole of NBS in Department of Commerce, Department Order No. 90, "National Bureau of Standards" (30 January 1964). The repetition suggests how central the Institute of Material Research was to the NBS mission.

[70] Rexmond C. Cochrane, *Measures for Progress* (Gaithersburg MD: National Bureau of Standards, 1966), p. 69.

[71] See Annual Report, Institute for Materials Research, 1965.

[72] See "Presentation by Dr. Gordon K. Teal before Secretary of Commerce John T. Conner," 8 March 1965, p. 2. Copy of original from files of Gordon Teal held in the archives of the IEEE Center for the History of Electrical Engineering.

[73] Teal cites his time there as one of his most enjoyable experiences [Interview 1991, p. 121].

[74] Interview 1991, pp. 124–127.

[75] "Gordon Teal's achievements while in Washington" copy of original from files of Gordon Teal held in the archives of the IEEE Center for the History of Electrical Engineering.

[76] The examples and quotation to follow are from Annual Reports of the Institute of Materials Research, 1965 through 1967.

[77] Annual Report, Institute for Materials Research, 1965, p. 85.

[78] Teal, "The NBS Matter & Materials Data & Standards Program," talk to Industrial Research Institute, Detroit MI, 11 October 1966, p. 9.

[79] Annual Report, Institute for Materials Research, 1966, p. 69.

[80] Teal, "Summary of G. K. Teal's Washington activities and accomplishments."

[81] Teal, "Technology and human potential," AAAS Retiring Vice-Presidential Address, delivered in Chicago on 28 December 1970.

[82] In addition to these, Teal is a Fellow of the Texas Academy of Science (1954), the American Association for the Advancement of Science (1959), the Institute of Electrical Engineers (1966), the Washington Academy of Science (1966), and the American Institute of Chemists (1968).

[83] Edwin Layton, "Mirror-image twins: The communities of science and technology in 19th century America," *Technology and Culture*, vol. 12, 1971, pp. 562–580.

CHAPTER 5

From Automatic Volume Control to the Stationmaster Antenna

Harold Alden Wheeler
and Applied Electronics

HAROLD ALDEN WHEELER

Figure1. Harold Alden Wheeler, born 10 May 1903, is a recipient of the IEEE Medal of Honor and a current resident of Ventura, California.

For the United States the year 1930 may be regarded as the beginning of the golden age of radio.[1] By that time some 600 stations were broadcasting, many of them around the clock, and 46 percent of all households had receivers. In that year Lowell Thomas established the lasting institution of the nightly news, delivered electronically. That was when Rudy Vallee started the first of the radio "variety shows" that soon became immensely popular. Serial drama and comedy were rapidly gaining listeners; "Amos 'n' Andy" first went on the air in 1928, "The Rise of the Goldbergs" in 1929, and "Death Valley Days" in the following year. And classical music, supplemented by instruction in music appreciation, was being broadcast regularly; 1930 saw live opera broadcasts, including one from Dresden, Germany, and the beginning of the regular transmission of Sunday concerts of the New York Philharmonic Orchestra.

It was, no doubt, the aficionados of classical music who were then particularly eager to purchase the most advanced Philco radio, the Screen Grid Plus, which had appeared on the market in the fall of 1929. This radio, known as the Model 95 (see Figure 2), boasted an unusually even response across the audio frequencies (giving "balanced" reception) and a new feature, diode AVC (automatic volume control).

Automatic volume control, which had earlier been available only in a more expensive and less effective form, was a remedy to two problems: the tendency of distant stations to fade and swell, and "blasting," which happened whenever one encountered a local station in tuning the radio. The Model 95 was a phenomenal success, and it helped make Philco the leading radio manufacturer in 1930 and throughout the decade.[2]

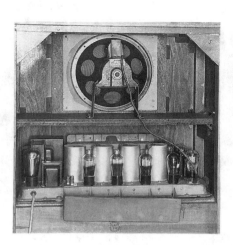

Figure 2. The chassis of the Philco Model 95, also known as the Screen Grid Plus, is shown on the left. It was sold in cabinets of varying style and quality; one of them is shown on the right.[3]

Just three years earlier Philco, the Philadelphia Storage Battery Company, appeared doomed: Its two principal products—batteries for radios and rectifiers to allow radios to be powered by house current—were suddenly rendered obsolete by the availability of AC tubes.[4] It was at that point that the company decided to manufacture radios and turned to a radio-engineering firm, Hazeltine Corporation, for expertise in a field new to Philco. The design of the Model 95, like the design of Philco's first radio, the Model 76, was largely the work of a young engineer at Hazeltine, who incorporated into the design his own invention of diode AVC. That young engineer was Harold Alden Wheeler.

Growing up in South Dakota and Washington, D.C.

Wheeler belongs to the generation of electronics engineers who grew up on crystal radios.[5] He was born in 1903, when Guglielmo Marconi, R. A. Fessenden,

John Stone Stone, and many others were working to improve and find a market for the infant technology of wireless transmission. Wheeler's father, William Archie Wheeler, taught agronomy at the South Dakota State College in Brookings until 1907, when he moved to Mitchell, South Dakota, to manage the Dakota Improved Seed Company, which he had helped found. In 1916 he accepted a position with the US Department of Agriculture in Washington, D.C. Wheeler's mother, née Harriet Maria Alden, was thought to be a descendent of John and Priscilla Alden, made famous by Longfellow.

As a boy, Wheeler was fascinated by two technologies just emerging in the first decade of the century: aviation and wireless communication. He remembers the awe evoked by wireless, as when an itinerant lecturer came to Mitchell to demonstrate radio signaling or when an older boy in town used a spark transmitter to steer a model boat. By the time the family moved to Washington in 1916, Wheeler, who had just then completed seventh grade, had decided on radio engineering as a career.

In a workshop in the garage (later the basement) of the family home, Wheeler built simple electrical devices, including an electric motor and galvanometers for measuring weak currents. He attached a crystal detector and headphones to bedsprings and a radiator pipe to listen in on Navy radio signals during the period of US involvement in World War I. After the war he continued his experiments, building several receivers and a transmitter. To qualify for an operator's license he attended evening classes at the National Radio Institute, and in April 1920 he received a license for an amateur station (see Figure 3). In November of that year, the station KDKA in Pittsburgh began regular broadcasting, and this greatly increased the attraction of radio as a hobby. In late 1921 Wheeler built an audio amplifier and a loudspeaker and thereafter often invited large groups of people to the house to listen to KDKA and other stations.[6]

His son's hobby alerted Archie Wheeler to the possibilities of using the new technology for the benefit of farmers. In 1920 he established, as a service of the Department of Agriculture, the Radio Market News Service, which brought the latest market reports to farmers.[7] (Figure 3 is actually a picture of Wheeler receiving the first broadcast of the Radio Market News Service on 15 December 1920.) The father's involvement with radio proved subsequently to be of great value to the son.

College Years

In 1921 Wheeler graduated from Central High School and, having won a scholarship by competitive examination, enrolled at George Washington University. This allowed him to live at home while attending college, which was a great advantage because, as he put it, "at home I had my basement laboratory where I spent more time than I did studying, and that was the beginning of my later career."[8] With his father's assistance, Wheeler obtained summer employment in 1921 and 1922 at the Radio Laboratory of the National Bureau of Standards (NBS), which was directed by John Howard Dellinger.[9] Here he gained experience with some of the most advanced techniques and devices in radio engineering and also became acquainted with many leaders of the new field, such as Frederick Kolster and John M. Miller.

Figure 3.   Wheeler and the apparatus for amateur station 3QK.

At that time one way people sought to improve radio reception was by placing a tuned radio-frequency (TRF) amplifier ahead of the detector. The major difficulty was that internal feedback often caused oscillations (heard as whistles or squeals) in the triode vacuum tube used as amplifier. In July 1922 Wheeler devised a circuit to neutralize the internal feedback, and a few days later he built and tested the circuit in his home laboratory.

This invention and a chance meeting the following month set the direction of Wheeler's career. He was accompanying his father on a business trip that in-cluded a visit to a seed company in Hoboken. At a restaurant there, they encountered Alan Hazeltine, professor at nearby Stevens Institute of Technology, who knew Archie Wheeler because both were members of the National Radio Conference, which was organized in 1921 by Secretary of Commerce Herbert Hoover to bring order to the chaos of radio broadcasting. Hazeltine invited the two of them to stop by his office at Stevens that afternoon.

At Hazeltine's office, Wheeler described his neutralized TRF amplifier. Hazeltine, greatly surprised, said that he had made the same invention about three years earlier—he showed Wheeler a pending patent application—but had not implemented it in practice. Two months later, at Hazeltine's suggestion, they discussed an arrangement for cooperation. Wheeler agreed to give Hazeltine rights to some related inventions in exchange for 10 percent of any royalties for the neutralization patent.[10] It was agreed further that Wheeler would work for Hazeltine the following summer. The connection between Hazeltine and Wheeler proved to be a lasting one.

Hazeltine joined the electrical engineering department of Stevens Institute of Technology in 1907 and became department head in 1917. (See Figure 4.)[11] Like Dellinger of the NBS Radio Lab, Hazeltine worked to make a science of radio engineering, as in his 1918 paper, "Oscillating audion circuits," which was the first mathematical treatment of oscillating vacuum-tube circuits.[12] During the war Hazeltine designed for the Navy the radio receiver SE-1420, which went into wide use. In 1922 a group of radio manufacturers asked Hazeltine to design a radio receiver that incorporated his invention of neutralization. These manufacturers were seeking a radio design that would not infringe Edwin Armstrong's patent on the regenerative tube detector, the rights to which were held by RCA. Hazeltine's receiver, called the "Neutrodyne," sold so well that the fourteen companies licensed to produce it could not keep up with orders.[13] It dominated the market until 1928, when the availability of screen-grid tubes for amplification made it obsolete.

In early 1924 Hazeltine established Hazeltine Corporation to manage the neutralization and other patents and to provide engineering services (mainly advice on product designs) to licensees. Wheeler, who had worked for Hazeltine the previous summer, was employed, as he puts it, "for all the time I could make available while continuing my college work."[14] Hazeltine became Wheeler's mentor, teaching him to take a theoretical and experimental approach to design problems. Hazeltine's style is expressed in his statement, "My inventions ... have all been the result of theoretical studies, verified and modified by subsequent experimental work."[15] Wheeler acknowledges a great debt to Hazeltine, saying, "From the time of our accidental meeting in 1922, he indoctrinated me in his imaginative and creative approach to any problem."[16]

In 1925 Wheeler completed a B.S. in physics and, as further preparation for a career in engineering, continued his study of physics at Johns Hopkins University. (Wheeler took his father's advice to study physics rather than electrical engineering on the grounds that EE education at the time was superficial in its treatment of the relevant science.)   In these years, beginning with summer employment in 1923, he was giving more and more of his time to his work for Hazeltine Corporation, and in 1928 he left Hopkins, without quite completing the requirements for a Ph.D., to work full-time for Hazeltine.

Figure 4.   Louis Alan Hazeltine (1886–1964), radio pioneer, professor at Stevens Institute of Technology, and founder of Hazeltine Corporation. He was elected Fellow of the IRE in 1921, Fellow of the AIEE in 1935, and President of the IRE for 1936.

In the three years at Johns Hopkins, Wheeler did important work on several projects not connected with Hazeltine Corporation. He assisted

Gregory Breit and Merle Tuve in 1925 in designing a "pulse transformer" that Breit and Tuve used in their famous measurement of the height of the ionosphere (by reflecting radio pulses off of it).[17] Wheeler made a theoretical study of wave filters composed of repeating sections.[18] He also worked with E. Cowles Andrus, a researcher at Johns Hopkins Hospital, to design and build a device for studying the refractory period of the heart (the period after the initiation of contraction during which the heart is not responsive to electrical stimulation).

Life at Hopkins was a very leisurely experience. They had a small dormitory for graduate students where I had two rooms that I shared with one of my friends. Meals were served inexpensively. So I was in direct contact with the graduate students, mostly in physics and chemistry. The quality of education, however, was not that great. They had qualified professors—some were famous—but the science of education hadn't permeated colleges yet. So the method of education was not terribly good. Nevertheless by attending Johns Hopkins I did gain a broader perspective on physics, at a leisurely pace.

. . . I was commuting to New York at least once a month and [working there] during summers. I was able easily to keep up with the pace and continue these outside activities.

Interestingly, one of these outside activities was the building of a pulse transformer. Among my colleagues and friends as graduate students were Breit and Tuve, who became famous in later years for their work with the ionosphere. [With] Tuve I had a particularly close affiliation because he also came from Minnesota. One evening after dinner they cornered me, and they said, "We've got a problem. We're making a transmitter for high-powered pulses to send up to the ionosphere and reflect back again. And we need a transformer, and we don't know how to make a pulse transformer." Well, during the summers I had worked with Hazeltine, one of the projects he was working on involved what was then unknown—a pulse transformer—and I was familiar with its workings. I translated that knowledge to Breit and Tuve's requirement, and I designed the first radar pulse transformer. It worked in their experiment.[19]

Automatic Volume Control

It was in the summer of 1925 that Wheeler made perhaps his most important invention, that of automatic volume control, also known as automatic gain control. As stated above, this circuit, incorporated in the Philco Model 95, alleviated the problems of "blasting" and fading. Wheeler described the invention in a 1928 article in the *Proceedings of the IRE* [20] (see Figure 5), and in the 1930s it became a standard feature of AM radios. It is still standard. The fiftieth anniversary edition of *Electronics* magazine, which appeared in 1980, contained a section entitled "Classic circuits," which presented twelve circuits that "have been basic to the commercialization of radio, TV, and computers and other data-handling systems." Wheeler's automatic volume control was one of the twelve.[21]

Diode AVC, though invented in 1925, was not immediately used in radio circuits because it was difficult to get much radio-frequency (RF) amplification ahead of the detector, and diode AVC functioned by using the level of the signal in the detector to control the gain in the RF amplifier. When the screen-grid tube, which was very effective as an RF amplifier, became available in 1928, this difficulty was removed. Thus it was that the Philco Model 95, which came out in 1929, was the first commercial radio to incorporate diode AVC.

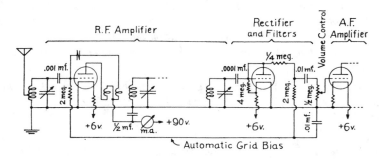

Figure 5. Diode AVC, the circuit Wheeler designed in 1925 for achieving automatic volume control in radio receivers.

The availability of the screen-grid tube had another effect: It rendered obsolete the neutralization that was at the heart of the Neutrodyne receiver. The only source of income for Hazeltine Corporation was the royalties it received from those licensed to use its patents; companies received a blanket license for all Hazeltine patents together with engineering services of Hazeltine engineers. Since the key patent in the Hazeltine portfolio had been the neutralization patent, it was fortunate for the company that just at that time the diode AVC patent became very important, helping to retain current licensees and attracting new ones.

In the summer of 1925, just after graduating from George Washington University in Washington D.C., I went to work a third summer in the Hazeltine Corporation laboratory. It was located in a few rooms in the attic of the Electrical Engineering building of Stevens Institute of Technology in Hoboken, New Jersey, across the Hudson River from New York City.

I was impressed with the amount of RF amplification in the latest designs of tunable receivers for radio broadcasting. Connected with an outdoor wire antenna, the sound level on local stations was so great that a "volume control" was needed to reduce the sound to a level suitable for listening. There were two tuning dials for antenna circuit and RF amplifier so a "third hand" was needed for the volume control. I perceived that an automatic control was needed to set the sound volume at a desired level for all stations. Then the volume control knob could be set once for all stations within range. I soon decided that the automatic control should be applied to the RF amplifier ahead of the detector and AF amplifier. I aimed to do that by developing from the detector a bias voltage sufficient to control the gain in the RF amplifier if applied to the control grid of one or more of the RF amplifier tubes.

It was customary to operate the detector with a rather small signal voltage because it was difficult to design a large amplification ahead of the detector in the TRF amplifier. I investigated various circuits for amplifying the small rectified voltage in the detector, but these circuits were complicated and required critical adjustments.

In the fall of 1925, I entered the graduate school in Johns Hopkins University in Baltimore, near my home in Washington. I planned to make a receiver with automatic volume control during the Christmas holidays in my basement laboratory at my home. I decided on a superheterodyne circuit to obtain sufficient RF amplification ahead of the detector. It was my first because that type of receiver was not in common use outside RCA (they refused to license its use by other companies). My design was an advance over others because I used a neutralized intermediate-frequency (IF) amplifier at a frequency only slightly below the broadcast band (550–1600 MHz). This was a preview of later practice when the superheterodyne went into universal use a few years later, in 1930.

I built this receiver on a work table. It was later preserved for exhibit in litigation so a photograph is available. It is now preserved in a showcase at the Hazeltine Corporation headquarters in Greenlawn on Long Island, New York. It incorporated a modular design on eight small wooden bases. I had accumulated the necessary parts before the holidays. There were tunable circuits for the antenna, local oscillator, and four IF circuits. The construction of the receiver took only a few days and its operation was spectacular.

Then I approached the problem of obtaining from the detector a bias voltage for automatically controlling the amplification in the first one or two of the IF stages. I had entered in my notebook some rather complicated circuits using the usual triode detector. One problem I recall was the unavailability of resistors of values interme- diate between wire-wound (up to 1000 ohms) and graphite "grid leaks" (around 1 megaohm). I needed some around 10 to 100 kiloohms. Also, those circuits required a floating "B battery" in addition to the common grounded B battery for all amplifier stages. For a few days, I was making little progress.

Then it occurred to me that I had enough amplification to use a diode rectifier as the detector, operating at 10 volts of IF signal. The diode would be made of the available triode with grid and plate connected together. This was heresy, to sacrifice the gain of one triode, costing about $5 apiece (about $100 in today's currency). However, the greater RF (IF) gain made up for the customary gain in a triode detector, so I still needed only two audio (AF) stages to drive the loud speaker. I quickly con- nected the diode with resistors and capacitors to provide separately the audio (AF) signal voltage and the bias (DC) voltage for control of the first one or two IF stages.

I got around to entering the theory of these connections in my notebook a few days later (January 2, 1926). This was the date we established later for "reduction to practice." [22]

Full-time at Hazeltine Corporation

When Wheeler began full-time employment for Hazeltine in 1928, his first assignment was to develop a standard signal generator for use in testing radio receivers. The device that he and a coworker built was far in advance of anything being used at the time. This was typical of Hazeltine Corporation, which then and in the 1930s led the industry in the development of test equipment. Indeed, the ability of Hazeltine engineers to measure the performance of circuits and components played a large role in the company's success. Wheeler has said, "I can't overemphasize the theme of testing equipment as being a real limitation on the rate of progress." [23]

This was an area to Wheeler's liking, partly, as he put it, "because you didn't have to worry about the market price as you did in receiver design." [24] His most important invention in this area was what he called the "piston attenuator." It was customary, in testing a radio receiver by means of a standard signal generator, to reduce the signal in steps using a ladder network of resistors. Wheeler succeeded in constructing a device that permitted continuous variation of the signal strength. Its name came from its cylindrical form and the fact that one electrode was moved along the axis of the cylinder. [25] This was a forerunner of propagation waveguides, which assumed great importance in the development of radar; indeed, the piston attenuator was reinvented during World War II at MIT's Radiation Laboratory, where it was called a "waveguide beyond cutoff."

Test equipment remained one of Wheeler's principal interests, as a list of his publications suggests. A device to make a photographic record of the frequency variation of the sound from a loudspeaker and a device to measure very small inductances are examples of his work in the 1930s. [26] In 1962 he was invited to write for the fiftieth anniversary volume of *Proceedings of the IRE* a review paper entitled "Microwave measurements." [27]

From the start of his full-time employment at Hazeltine in 1928 until World War II, Wheeler continued work on radio circuits. In the 1930s a principal aim of radio engineers was to lower the price of receivers; Hazeltine engineers, by designing inexpensive superheterodyne receivers, led the industry in this area. Hazeltine excelled also in evaluating loudspeakers—as part of the 1930s movement toward "high fidelity"—and in designing so-called "all-wave receivers" (which could also receive short-wave transmissions from Europe) and, at the end of the decade, FM receivers. In addition, Wheeler says, "we developed some extremely sophisticated broadcast receivers with automatic controls, most of which never reached the commercial market." [28] For use with diode AVC, Wheeler invented a tuning meter, which showed the change of amplification effected by the AVC and which is named for its usefulness as an aid in tuning; various forms of it soon appeared in radio receivers, one being the RCA "Magic Eye."[29]

Wheeler attributes his success in radio design to his taking a scientific approach to problems, which came from his training in physics and the indoctrination in this approach he got from Hazeltine. By the time of US entry into World War II, Wheeler had received more than a hundred US patents. In addition, he published a number of important papers on radio receivers; next in importance to his 1928 paper on diode AVC was probably the lengthy "Theory and operation of tuned radio-frequency coupling systems" coauthored by William A. MacDonald, chief engineer of Hazeltine.[30] Several quite theoretical papers he wrote on FM reception were influential.[31]

Patent Litigation

The story of the development of radio can hardly be told without considerable discussion of patent litigation. Lee De Forest, Edwin Armstrong, and David Sarnoff—three of the most prominent radio pioneers—were all involved in lengthy court battles. The uncertainty of patent rights in the United States no doubt stimulated both infringement and litigation and probably slowed technical progress.[32] In the 1920s, Hazeltine Corporation defended one of its licensees (Garod Corporation) against a suit brought by RCA, charging that the Neutrodyne receiver infringed two patents owned by RCA, and Hazeltine Corporation brought suit against a number of companies for infringing its neutralization patent. Wheeler often served as an expert witness in these cases.

In the 1930s there were a number of cases that concerned Wheeler's patent on diode AVC.[33] Many radio manufacturers were licensees of Hazeltine, so had unquestioned right to use the invention. A number of companies, including RCA and Detrola, used it without taking a license. First, Hazeltine Corporation sued RCA for infringement. After a full trial, but before a decision was given, RCA agreed to take a license. Hazeltine then sued Detrola for infringement. A Federal district court ruled the patent valid and infringed, and this decision was unanimously upheld by the circuit court of appeals. Detrola then appealed to the Supreme Court. In a unanimous decision in 1941, the Supreme Court reversed the verdict of the lower courts, saying, "We conclude that Wheeler accomplished an old result by a combination of means which, singly or in similar combination, were disclosed by the prior art and that . . . he was not in fact the first inventor, since his advance over the prior art, if any, required only the exercise of the skill of the art."[34]

Wheeler explains the adverse decision as resulting largely from the failure of the Supreme Court to make a careful study of this case (suggested by the unanimity of the decision) and its long-term bias against enforcing patent rights. In the first half of this century, partly because patents were seen as monopolistic devices, it became more and more difficult to obtain a court ruling that a patent was valid and had been infringed. Of all patent cases brought before courts of appeal and the Supreme Court in the period 1941 to 1945, 89 percent were declared invalid,[35] and in 1949 Supreme Court Justice Robert H. Jackson said, only partly in jest, "the only patent that is valid is one which this Court has not been able to get its hands on."[36] That most engineers did not agree with the Court is shown by the fact that many companies, including GE and RCA, had agreed to pay royalties for the use of the invention, and by the recognition that professionals have accorded Wheeler as the inventor of automatic volume control.[37]

Because the diode AVC patent was the most important in the Hazeltine portfolio of patents and because the only source of income for Hazeltine Corporation was royalties, the Supreme Court's decision was a devastating blow to the company. As it happened, however, the coming of World War II allowed the company to turn in a new direction—defense contracting. After the war, the adverse climate for patent rights and the demand for military electronics caused Hazeltine to change its mode of operation: Instead of earning money only through patent licensing (including engineering support), the company expanded by seeking and winning defense contracts.

Wheeler was involved in another patent dispute that reached the Supreme Court. It occurred much later in his career and concerned television rather than radio, but it resembled the first case in that the Supreme Court reversed the decision of a circuit court and in that Hazeltine Corporation suffered a crippling blow. Zenith Corporation, sued by Hazeltine for patent infringement, made an antitrust counterclaim, contending that Hazeltine's foreign licensing practices denied Zenith certain export markets. A judgment in the US District Court in Chicago that Hazeltine pay Zenith damages was reversed by a circuit court but reaffirmed by the Supreme Court. As in the earlier case, the fact that the decision by the Supreme Court was unanimous suggests to Wheeler that the judges did not give the case careful consideration.[38]

Television Engineering

It was largely in the 1930s that the technology of television was developed. In the United States, RCA, with its star engineer Vladimir Zworykin, was the acknowledged leader, and in early 1933 RCA built and operated an entirely electronic television system.[39] A year earlier Hazeltine Corporation began work on television receivers, and in the years 1936 to 1940

Hazeltine engineers built a series of television receivers of increasing sophistication.

Wheeler gave particular attention to two problems: circuits for scanning the picture tube, and circuits for amplifying the wideband signals that television required. He carried out a largely theoretical study of the reproduction of an image by the scanning process. A product of this work was a paper he coauthored with Arthur Loughren, "The fine structure of television images," which became a basic reference on the subject.[40] Wheeler's study of distortion in a wideband amplifier led to the introduction of the concept of "paired echoes" and the publication of the article, "The interpretation of amplitude and phase distortion in terms of paired echoes," which is probably Wheeler's most famous paper.[41]

Television, in contrast to radio, operates with signal pulses rather than a continuous carrier. In order for an amplifier in a television receiver to reproduce pulses accurately it must both respond to the wide range of frequencies making up a pulse and maintain the phase relations of the components. The latter requirement necessitates the inclusion in amplifying circuits of phase correction. The phase distortion could be described in terms of an "echo" pulse before and after the main pulse. Wheeler introduced the term "paired echoes" and showed that any phase distortion can be analyzed as made up of a symmetrical pair of echoes and an asymmetrical pair.

Wheeler went on to make a general study of wideband amplification and introduced a number of other concepts—such as "dead-end filter," "feedback filter," and "speed of amplification"—useful in the analysis of wideband amplification. Some of this work is reported in a very influential paper published in 1939, "Wide-band amplifiers for television."[42] This and other papers brought him in 1940 the Morris N. Liebmann Award of the IRE, "For his contribution to the analysis of wideband high-frequency circuits particularly suitable for television."

World War II

World War II caused Hazeltine Corporation to set aside television development, but the experience gained with pulse techniques—pulses of current produce the spots of which the television image is composed—proved valuable in work on radar. After the war Wheeler did not return to television circuits, but Hazeltine Corporation became a leader in the development of color television.

In June 1940, as US involvement in the European war was appearing increasingly likely, President Roosevelt approved a plan advanced by Vannevar Bush to coordinate civilian research for military ends. The coordinating agency, the National Defense Research Committee, arranged for the participation of civilian scientists and engineers in military research and development. Hazeltine Corporation, which became involved in several military projects, was given sole responsibility to develop, for the Army Corps of Engineers, a device to detect buried antitank mines.

Wheeler, who directed this project, started with an existing device, a so-called "treasure finder." In Wheeler's words, "The treasure-finder was some coils on a long pole, and the coupling between the coils had to be critically balanced out in order to detect the reaction of a buried metal object. Their principal defect was that the balancing was so critical that even exposure to daylight would cause expansion and upset the balance."[43] Wheeler replaced the double coil with three coplanar concentric coils, the opposing inner and outer coils being the transmitter and the intermediate coil the receiver. As his calculations had predicted, the balance between the coils became insensitive to temperature change. This innovation, together with an inspired mechanical design that was largely the work of Hazeltine engineer Leslie Curtis, yielded an effective mine detector, the SCR-625 (see Figure 6), which was rushed into service for the North African campaign of 1942.[44] The SCR-625 went into quantity production (by another company) and saw wide use in World War II and in the Korean War.

Among the other projects in which Hazeltine was involved was an exploration of the possibility of a television-guided bomb. Though the feasibility of the idea was demonstrated, the project with Hazeltine was discontinued. It soon happened, however, that work on the development of radar, first for the Signal Corps and then for the Navy, required all of Hazeltine's resources, so no other projects were undertaken.

A technology in the early stages of development before the war, radar was of such obvious military value that the major combatants soon developed it to the point of being able to detect airplanes and ships at great distances and through clouds and fog. But the blip on the radar screen did not reveal whether or not the airplane or ship was hostile. IFF, Identification Friend or Foe, was a system designed to answer this question. Friendly aircraft and sea vessels were equipped with a transponder or radar beacon, which, when triggered by a surveillance radar (or by a separate transmitter called an interrogator), sent out a coded reply. The absence of a reply was taken to mean that the airplane or sea vessel was hostile. The British tested an IFF system, the Mark I, in 1939; Mark II and Mark III were improved British systems.[45]

Hazeltine obtained the contract to design all the Navy's IFF equipment and arrange for its manufacture. Starting with an experimental model of the Mark III, Hazeltine engineers designed an improved system. Because IFF used a frequency band higher than Hazeltine had dealt with before, Wheeler undertook studies of the antennas and transmission-line circuits appropriate to this frequency range.

Earlier, Wheeler had done work in antenna design. In the 1930s, as receivers were designed for radio waves of higher and higher frequency (6 to 18 MHz), it became possible to make great improvements in antennas. Thus, says Wheeler, "antenna design became a new field of expertise which I embraced immediately."[46] Because the IFF system used an even higher frequency range (157 to 187 MHz), new antennas were required. Wheeler set up within Hazeltine an antenna group, and over the next several years this group of three or four engineers, under his direction, designed a series of IFF antennas—for aircraft, surface vessels, submarines, and ground

Figure 6.   The mine detector SCR-625, widely used in World War II and the Korean War, designed by Wheeler and other engineers of Hazeltine Corporation.

stations. Most famous was the so-called "lifesaver antenna," which contained three radial spokes in a wheel beneath a vertical monopole antenna (see Figure 7); by the end of the war, this antenna had been placed on all of the Allied ships. Other antennas he designed were a vertical antenna with folding legs (for the paratrooper beacon PPN-1 used to guide airborne landings behind enemy lines in Italy and France) and a horizontal half-loop antenna (for transponder beacons, such as the Signal Corps YH beacon used in the Pacific). After the war, the Navy recognized Wheeler's contributions by awarding him a Certificate of Commendation.

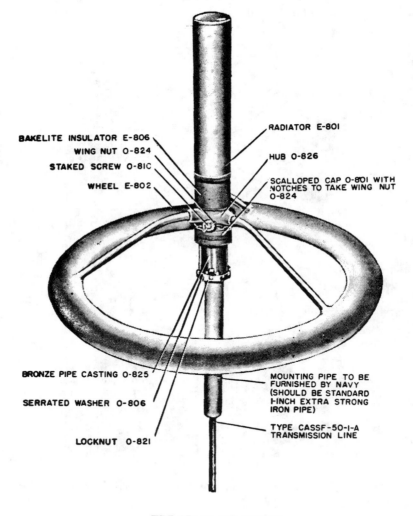

TYPE CTZ-66-AFJ ANTENNA.

Figure 7. The antenna CTZ-66-AFJ, known as the "lifesaver antenna," widely used in World War II, designed by Wheeler and other engineers of Hazeltine Corporation.

WHEELER: *... about the middle of the war our IFF system had been sort of compromised, and the need was recognized for a more advanced system. So in parallel with our immediate needs, a project was initiated by the Navy as a leader of all agencies, both in the government and civilian, to work on a successor IFF. After the Mark III there was one designated Mark IV that didn't go anywhere. The next was designated Mark V, and that is the one that was the subject of intensive developmental activity in the latter half of the war period.*

The Mark V was developed under cognizance of the Naval Research Laboratory located in Washington, D.C. That project involved cooperation of all the government military agencies and several companies with ambitions to manufacture the equipment. It was designated the Combined Research Group (CRG) with headquarters in a new building at the Naval Research Laboratory. I and our group spent a great deal of time on location. Our responsibility in the new project—the immediate responsibility—was to provide by a quick reaction new equipment that they perceived they needed. And so we had a shuttle between our headquarters in Little Neck and NRL in support of that activity.

... Now this CRG design designated Mark V was mainly distinguished by a higher frequency band, the so-called L-band, around 1 megahertz. So that involved a new set of technical problems. In the antenna area and in the area of high frequency circuits, I was a leader. So I have some patents in that field. That project came to a screeching halt when the war was over, by which I mean the ambitious program was reduced to a holding operation. It shortly resulted in a new simplified IFF, which became the keystone of our postwar IFF equipment. It was designated Mark X.

You would naturally be curious what happened to the intervening numbers. One of the leaders in the government, Gene Fubini, was at the blackboard one day and talking about the next generation of equipment. He wrote down Mark X [the letter X, indicating an unknown number], and that was transcribed to Mark Ten.

NEBEKER: *So there were no VI, VII, VIII, and IX?*

WHEELER: *None. And Hazeltine became the leader in both development and manufacture of Mark X and successor equipment.*[47]

Wheeler Laboratories

In the immediate postwar years there were exceptional opportunities for electronics engineers. Because he was well known, Wheeler knew that he would have no difficulty working as a consulting engineer and decided to leave Hazeltine Corporation. One of the first people to contact him was Robert Poole, head of the Whippany laboratory of Bell Laboratories.

In 1926 Bell Laboratories had established a small research facility at Whippany, New Jersey, about 30 miles west of New York City. The rural setting of this laboratory made it the choice for the site of top-secret research on radar that Bell Laboratories agreed to do for the Navy, and in 1938 a new engineering group was set up at Whippany.[48] After the war, the Whippany Laboratory was the site of the development of the electronic guidance equipment and other components of the Nike antiaircraft missiles.[49]

Wheeler, working with two other engineers, obtained a subcontract to design microwave circuits for the Nike System. In 1947 Wheeler Laboratories was incorporated, and the number of employees grew steadily over the next decade, most of the work being subcontracts from the Whippany Laboratory. Development of the Nike Hercules System began in 1953, and for this project Wheeler Laboratories designed microwave circuits and an innovative target-tracking antenna—a double-reflector antenna that is still in use today.[50] Wheeler said of his company, "... we enjoyed the luxury of having the primary management handled by Bell Laboratories, and we could devote almost all of our attention to creative work."[51]

Antenna design remained an area of special interest to Wheeler. He carried out a theoretical investigation of "small antennas"—antennas of dimensions much less than the wavelengths they are designed to receive—that revealed simple relationships and rules. Over the next several decades he published papers both on principles of antenna design and on particular antennas whose design benefited from the theoretical studies.[52]

For example, when consulted about the small antenna for a proximity fuse placed in the nose of a missile, he proposed a design that was much more efficient than earlier ones. As part of his work for a Navy project to use VLF (very low frequency) waves to communicate with submarines, he helped design the world's largest antenna, which was, according to the definition given above, a small antenna. Built at Cutler, Maine, and commissioned in 1961, this antenna was supported by 26 towers, each about 1000 feet tall, and extended over two square miles. A second such antenna, supported by 13 towers, was built in Australia (see Figure 8).[53]

In pondering antenna design for low-frequency radiation where the ground was one electrode of the antenna, Wheeler perceived that the earth's crust could be regarded as a parallel-plate waveguide. The surface layer was conducting because of moisture; the layer below that was nonconducting, being dry and consisting largely of silica; and at a greater depth the high temperature made the silica conducting. At a conference in 1950 Wheeler presented his idea, pointing out that this might provide "a channel for radio waves to go long distances in the earth's crust and then come out deep in the ocean...."[54] Because the Navy was searching for ways to communicate with submerged submarines, a secrecy order was placed on his patent and a project—the Navy's ELF (extra low frequency) project—was begun to investigate its possibilities.

During World War II there were some radar antennas that consisted of a small number, perhaps half a dozen, of separate radiators. These so-called array antennas were steered mechanically. A number of people began thinking about electronic steering of the radar beam, one of whom, Arthur Loughren, was at Hazeltine Corporation. Loughren devised a type of steering called frequency steering; others later devised ways to steer the radar beam by varying the phase of the radiation, and so-called "phased arrays" became widely used.

Figure 8.   The U.S. Navy megawatt VLF antenna built at North West Cape, Australia.

Wheeler also became interested in array radars and published in 1948 one of his most influential papers. Entitled, "Radiation resistance of an antenna in an infinite array or waveguide,"[55] it helped to establish a scientific basis for the design of array elements. Wheeler proposed that, in the design process, the array element be treated as if it were an element of an infinite array. The idea found immediate application in work under-taken by Wheeler Laboratories for Bell Laboratories. In the 1960s Wheeler published several other papers on phased-array antennas.

Wheeler specialized also in the theory and design of transmission lines, especially strip lines, which are thin conducting strips separated from a return conductor (either another strip or a ground plane) by a nonconduct-ing sheet. Although the idea of printed circuits goes back to the turn of the century, they did not begin to be manufactured in quantity until the 1950s. The properties of strip lines, or "microstrips," were difficult to calculate, and Wheeler made important contributions in this area, especially in the 1960s and 1970s. His paper "Transmission-line properties of parallel strips separated by a dielectric sheet," published in 1965, was named a Citation Classic by *Current Contents*.[56]

Both in Hazeltine Corporation and in his own company, Wheeler gave much attention to the continuing education of the engineers. At the laboratory of Hazeltine Corporation in Bayside, New York, in the 1930s, Wheeler organized classes and did most of the teaching himself. He did the same at Wheeler Laboratories. He also continued the Hazeltine practice of having engineers write reports, both on improvements they had made and on tests performed on a licensee's design. The resulting reports, like the ones issued by Hazeltine engineers earlier, were highly valued by the clients.

Wheeler's scientific approach to engineering problems often resulted in lengthy manuscripts. Because *Proceedings of the IRE* contained only articles of moderate length, Wheeler turned to private publication. In the years 1948 to 1954, nineteen articles, some of them more than 40 pages long, appeared in the series entitled *Wheeler Monographs*, published by Wheeler Laboratories. These were later collected in two bound volumes.

Back to Hazeltine Corporation

In the 1950s, Wheeler Laboratories prospered, and in 1959 it employed about a hundred engineers. The military had been spending a great deal of money on technological innovation, but in 1959 such money became scarcer and Bell Laboratories had less need to subcontract work to Wheeler Laboratories. This circumstance and the friendly relations Wheeler had maintained with Hazeltine management made him receptive to an offer by Hazeltine Corporation to acquire Wheeler Laboratories. The acquisition took place in 1959, and from then until late 1970 Wheeler Laboratories remained a separate part of Hazeltine Corporation. In that year it was merged with the research laboratories of the parent company. Wheeler continued as president of Wheeler Laboratories until 1968.

. . . The '50s was a period that we'll never see again and never saw before, when the Pentagon was spending all the money they could on innovation. That wasn't the name of the game in the government generally. But that period after the war, which meant especially the early days of guided missiles, was a period unprecedented and that'll never happen again. So that was a fertile field for innovation. But then at the end of that decade, in 1959, happened what you wouldn't have thought possible. The Air Force sent a letter to the contractors which said money was getting scarce, and they couldn't expect to be paid on time.

Well, that was one of many events which started to inhibit the full range of innovative activity. It among other things reduced the Bell Laboratories load to a point where they gradually had less need for our services. I might say that from some viewpoint their contracting our services should have been frowned on by their organization because essentially they were building up engineering talent to compete with their own group. But in the climate of the '50s, that was not an important consideration. Later on the people who had introduced us in their group became less involved—either higher management positions or retirement. The successor group was less personally involved in our activities, and the work for them was tapering off.

It was just at that time that MacDonald came over to see me. He said, "Shouldn't we get together again?" Well, the circumstances left me receptive to that approach, so Hazeltine acquired our company at a nominal price in stock and very thoughtfully continued our activities for another decade, semi-independent. We still operated under our name, and we still had opportunities for contracts with various other organizations.[57]

In the 1960s, subcontracting work, mainly for Bell Laboratories, continued. Wheeler and his associates made designs for antiballistic missiles (ABMs) and for ABM radar. Especially interesting was some work Wheeler did about 1960 for the Communications Products Company. The task was to design an antenna, suitable for base-station communications, that would concentrate the radiation in the horizontal plane by means of a vertical array of vertical dipoles. Drawing on his theoretical studies of antennas, Wheeler designed what became known as the Stationmaster antenna, more than a quarter-million of which have since been sold.

One of the projects on which Wheeler worked in the 1970s aimed at the design of a radar antenna that would "interrogate" the altitude-coded radar beacons carried by aircraft. This antenna had to be light in weight and be subject to very little wind loading, since it was to be mounted on the rotating antennas of the surveillance radars already in operation at airports. Wheeler started with an open, flat array of many elements and then discovered a way to reduce the required number of rods by one half. The design was a great success and can be seen at major airports today in what looks like a bedspring on top of a rotating radar.

Wheeler was employed by Hazeltine Corporation from 1924 until 1987, except for 13 years after the war. In 1958 Alan Hazeltine wrote of Wheeler, "His brilliant inventions in many branches of electronics and his training of younger engineers were of great importance to the success of the company."[58] To this must be added mention of Wheeler's contributions to the management of the company; besides directing the company's Bayside Laboratory in the 1930s and serving as chairman of the Hazeltine Board of Directors from 1965 to 1977, he was named chief executive officer at the time of the crisis brought on in 1965 by the unfavorable outcome of the Zenith litigation. He took bold actions to restore employee morale, assisted in a reorganization of the company, and in 1966 turned over management to a new chief executive, David Westermann. Wheeler stayed on at Hazeltine past the usual retirement age in the capacity of chief scientist, providing support to many of the company's projects. He worked full-time until the early 1980s, when he reached the age of 80, then worked three days a week until 1987.

Handbooks and Formulas

Since his high school days Wheeler has had a hobby of collecting formulas. Whenever he encountered a formula that he thought might be useful, he entered it in a notebook. Before any "radio-engineer's handbook" appeared, there were collections of formulas at the back of catalogs of radio components (included so that people would keep the catalogs at hand), and Wheeler remembers these as one of his sources.[59]

Like most practicing engineers, Wheeler is grateful to the compilers of handbooks. One was his first boss, John Howard Dellinger, whose *NBS Circular 74*, entitled "Radio instruments and measurements," was the standard authority for two decades after its first appearance in 1918.[60] Another was Frederick Terman, whose *Radio Engineers' Handbook* became the vade mecum for radio engineers in the post-World War II decades. In the mid-1930s, when Terman began gathering material for a new handbook, he traveled to Hazeltine's Bayside Laboratory to talk with Wheeler, which was the first of many visits between the two engineers. When the handbook was finally completed in 1943, it drew on the work of some 900 authors; Wheeler was cited more often than any of them except George H. Brown (known for his work on antennas) and Terman himself.[61]

Wheeler has always argued for understanding design principles and practices and objected to carrying out computations by the use of a "black box," such as a canned computer program or—what was common throughout most of his career—a numerical table. He sought instead a perspicuous symbolic or graphical presentation of the quantitative relationships, in order that someone using the formula or graph could gain understanding of the physical phenomena, and he preferred analysis or synthesis at what he calls the "slide-rule level," rather than at a computer level. For his own work he did adopt computer methods in the 1970s, but reports jokingly that "The saddest moment of my life was when I put away my slide rule."[62]

Wheeler, who advocated the use of "every possible shortcut to relieve the engineer of unproductive nuisance in thinking, computing, and testing,"[63] published many articles presenting design formulas, charts and other calculating aids, procedures for the slide rule, and "graphical presentation of design relations in a form suitable for analysis or synthesis."[64] His first such publication, entitled "Simple inductance formulas for radio coils," appeared in 1928.[65] Others appeared sporadically in the decades that followed, and two appeared 54 years later.[66]

For example, in 1942 Wheeler published "Formulas for the skin effect."[67] It had long been known that a so-called "skin effect" reduced the effective cross-section of a conductor carrying alternating current, and thorough mathematical treatments had been written. What Wheeler did in this article was to describe the effect in more understandable terms, deriving from this description a simple way—called the "incremental-inductance rule"—of computing the skin effect in wires of various cross sections. This rule came into wide use for the design of transmission lines, especially strip lines.

Another example is provided by the formulas in Figure 9, which are useful in analyzing and designing printed circuits. They concern the properties of a strip line (or microstrip), defined to be a conducting strip separated from a parallel ground plane by a dielectric sheet. In an article published in 1977 (almost fifty years after his first published paper), Wheeler showed that a single equation was valid (within two percent relative error) for any strip width and any dielectric constant. Moreover, the equation is reversible, which makes it easy to use it for either analysis or synthesis.[68]

A 1946 editorial entitled "The real economy in engineering," which Wheeler wrote for *Proceedings of the IRE*, contains the following statements: "The real economy in engineering is the best use of every available aid in arriving at an understanding of the problem and an expeditious solution. . . . [understanding a problem] requires concentration, enlisting the aid of th e best references and charts, practicing on examples, developing new and simpler formulas and charts, outlining the limitations. . . . The pioneering in science will always leave in its wake a great demand for reference material which will aid in the solution of engineering problems by reducing the labor of computation and especially by doing this in such a way as to contribute to the understanding of the solution and its limitations."[69]

$$R = \frac{42.4}{\sqrt{k+1}} \ln\left\{ 1 + \left(\frac{4h}{w'}\right)\left[\left(\frac{14+8/k}{11}\right)\left(\frac{4h}{w'}\right)\right.\right.$$

$$\left.\left. + \sqrt{\left(\frac{14+8/k}{11}\right)^2 \left(\frac{4h}{w'}\right)^2 + \frac{1+1/k}{2}\pi^2}\right]\right\}$$

$$w'/h = 8\frac{\sqrt{\left[\exp\left(\frac{R}{42.4}\sqrt{k+1}\right) - 1\right]\frac{7+4/k}{11} + \frac{1+1/k}{0.81}}}{\left[\exp\left(\frac{R}{42.4}\sqrt{k+1}\right) - 1\right]}$$

Figure 9. The wave resistance R (or characteristic impedance) of a strip line on a dielectric sheet above a ground plane can be computed using the formula at the left, where w' is the effective width of the strip, h is the thickness of the dielectric sheet, and k is the dielectric constant. The mathematically equivalent formula below is useful in designing a strip line that is to have a given wave resistance.

Professional Activities, Family Life, and Historical Writing

Wheeler counts himself fortunate in having worked almost all of his career in the vicinity of New York City, which was the focal point of the radio profession. Offices and laboratories of RCA, Bell Telephone, ITT, and other companies were located in the area; important electronics research was done at Columbia University, City College of New York, and Polytechnic Institute; the Radio Club of America and the Institute of Radio Engineers were both based in the city. Wheeler attended meetings of the Radio Club, sometimes presenting his own work; in 1936 he was named Fellow, and in 1964 he received the highest honor of the Radio Club, the Armstrong Medal. Especially during the period he was working on circuit theory, he was also an active member of the American Institute of Electrical Engineers.

IRE activities, especially the meetings of the New York Section and the annual meeting, he found particularly valuable. Wheeler, who was named an IRE Fellow in 1935, served on several committees, notably the Technical Committee on Radio Receivers and the Standards Committee, for both of which he served as chairman.[70] From 1940 to 1945 he was a director of the IRE, and after the war he helped establish a Long Island subsection. During this time he proposed that all members be assigned to "Sections," geographically defined, and that the annual directory include biographies of the Fellows. These practices were adopted by the IRE and have been continued by the IEEE.

The family Wheeler started when he married Ruth Gregory (shortly after her graduation from George Washington University in 1926) is a great source of pride to him. Harold and Ruth, with their three children, Dorothy, Caroline, and Alden Gregory, formed a close-knit and harmonious family (see Figure 10). Wheeler had the advantage himself of having grown up in such a family. He writes: "Our father and mother were a wonderful team and our family relations were as nearly ideal as could be imagined. Kindness and love, respect and admiration, were the cornerstones. There was a spirit of cooperation and enough discipline. Good habits and cleanliness were taken for granted, by the example of our parents" and "The family built by my parents proved to be a remarkable accomplishment. . . . It offered us children a generous ration of nurture, protection, confidence, ambition and opportunity. With the strength of our family ties, none of us is ever alone."[71]

Wheeler reminds one of the Enlightenment rationalists in his efforts to reshape traditional practices according to reasoned design. He was a proponent of the MKS (meter-kilogram-second) system of units as preferable to the CGS (centimeter-gram-second) system.[72] He has championed what he calls the "logical date code," in which the temporal units are given

Figure 10. Dorothy, Harold, Alden, Ruth, and Caroline Wheeler in 1936.

in decreasing order: 1934 DEC 23, or 34 12 23, or, in Wheeler's usual style, 341223.[73] He has argued for including journal name, volume number, and date on each page of an article so that photocopies would automatically contain that information. He has suggested innovative ways to refer to sources and given thought to the best form of the many abbreviations used in technical writing.[74]

The honors accorded Wheeler include, in addition to those already mentioned, a twenty-dollar gold piece won as a paperboy in Mitchell, South Dakota, honorary doctorates from George Washington University, the Stevens Institute of Technology, and Polytechnic University, election to the National Academy of Engineering, the Pioneer Citation of the Radio Club of America, and the Microwave Career Award of the Microwave Theory and Techniques Society. Wheeler has been an advisor to the Department of Defense—he served on the Guided Missile Committee from 1950 to 1953 and on the Defense Science Board from 1961 to 1964. In 1964 he received the most prestigious award of the Institute of Electrical and Electronics Engineers, the Medal of Honor, "For his analyses of the fundamental limitations on the resolution in television systems and on wide-band amplifiers, and for his basic contributions to the theory and development of antennas, microwave elements, circuits, and receivers."

Wheeler has written three historical books, *Hazeltine the Professor* (1978), *The Early Days of Wheeler and Hazeltine Corporation—Profiles in Radio and Electronics* (1982), and *Hazeltine Corporation in World War II* (1993). The first of these Wheeler felt a responsibility to write in order to make known the great impact Louis Alan Hazeltine had on electronics engineering, an impact that a person judging solely by inventions and publications would not suspect. The second book is largely autobiographical and deals with the period up to World War II. The third book concerns the activities of Wheeler and of Hazeltine Corporation during the war.

NEBEKER: *You've also kept a diary much of your life.*

WHEELER: *By accident, but very fortunately, someone presented me with a little pocket diary in my high school years for Christmas. I started very casually to jot down things I was doing each day. I was enough interested to continue the next Christmas and bought a diary for myself. I engaged in entries more and more over the years until I graduated to a normal-size diary, bound volume. It is interesting that I did not do what the stories in the literature describe about diaries. My diary was not an exposition of my current emotional problems and other things, just a fragmentary entry of things I did. Sometimes I have a hard time remembering what it was when I read the diary. But I continued that so I have a complete set of diaries that track my daily activities pretty much from high school to the present time.*

NEBEKER: *Without large gaps?*

WHEELER: *One year I just didn't bother to get a new diary. I've regretted it ever since. So one year was missing. I think it was 1930. That convinced me that I shouldn't allow gaps.*

NEBEKER: *You used these diaries in your historical writings?*

WHEELER: *Oh, very much. And in matters of personal interest. What year did I take a first trip to the West Coast? What was the airport when I began to fly west? When I arrived in L.A., the airport was Burbank. Just a friendly little shack. So that's one of my treasures that I keep within reach all the time when I'm working.*[75]

In looking back over his seven decades of work with radio receivers (AM, FM, and shortwave), test equipment, television, radar, transmission lines, and antennas, Wheeler has commented that his fields of specialization "progressed with the growth of technology in my profession. . . . "[76] He appreciates his good fortune to have been born at about the same time as the field of electronics, for it was only in the early decades, he points out, that the learning rate of a professional could keep up with the growth of the field.

---

[1] George H. Douglas, *The Early Days of Radio Broadcasting* (Jefferson NC: McFarland & Company, 1987), and Christopher H. Sterling and John M. Kittross, *Stay Tuned: A Concise History of American Broadcasting*, second edition (Belmont CA: Wadsworth Publishing Company, 1990).

[2] Alan Douglas, *Radio Manufacturers of the 1920's*, vol. 2 (Vestal NY: Vestal Press, 1989), p. 231. If a Philco advertisement (in *Collier's*, 14 September 1929, reproduced in Douglas, p. 237) can be taken at face value, then the roughly simultaneous appearance of the Model 95 and of regular symphonic broadcasts was not merely coincidental. According to this advertisement, the officials of the Philadelphia Orchestra, directed by Leopold Stokowski, "steadfastly refused until the present time to send its glorious music out over the air," but that the level of radio reception had been raised so much, especially by Philco, that "Stokowski has consented to go on the air for the first time." (In the following year Stokowski began a 10-year collaboration with scientists and engineers of Bell Laboratories for improving sound reproduction; see the article by Robert E. McGinn in *Technology and Culture*, vol. 24, 1983, pp. 38–75.)

[3] Photos courtesy of Alan S. Douglas.

[4] Douglas, *Radio Manufacturers of the 1920's*, p. 231.

[5] The information about Harold Wheeler contained in this article comes mainly from the following sources: (1) an extensive oral history interview of Wheeler conducted by the author 29–31 July 1991 (from which an edited transcript has been prepared); (2) Wheeler's books *Hazeltine the Professor* (Greenlawn NY: Hazeltine Corporation, 1978) and *The Early Days of Wheeler and Hazeltine Corporation—Profiles in Radio and Electronics* (Greenlawn NY: Hazeltine Corporation, 1982); (3) Wheeler's other published writings; (4) copies of personal papers provided to the author by

Wheeler; and (5) a manuscript autobiography prepared in 1990 for Project LMS (Life Members' Stories), sponsored by the IEEE Life Member Fund Committee. (Items 1, 4, and 5, as well as a full list of Wheeler's publications, are available at the IEEE Center for the History of Electrical Engineering.) An earlier version of this article appeared in *Proceedings of the IEEE*, vol. 80, 1992, pp. 1223–1236.

[6] Wheeler used a regenerative detector with 2-stage audio amplification and a single Baldwin receiver connected to the horn of a phonograph.

[7] At that time radio was already being used to bring to farmers another type of information that needed to be disseminated rapidly—weather forecasts. In 1917 Earle M. Terry at the University of Wisconsin began transmitting weather bulletins to farmers (Douglas, *The Early Days of Radio Broadcasting*, p. 83).

[8] Interview 1991, p. 17.

[9] Dellinger, a Princeton physics Ph.D. and president of the Institute of Radio Engineers in 1925, contributed greatly to the establishment of the science of radio engineering. His "Principles of radio transmission and reception with antenna and coil aerials," read at a meeting of the American Institute of Electrical Engineers in 1919, was a landmark paper (see Donald McNicol, *Radio's Conquest of Space* [New York: Murray Hill Books, 1946], pp. 154–155). A rich source of information about Dellinger is Wilbert Snyder and Charles Bragaw's *Achievement in Radio: Seventy Years of Radio Science, Technology, Standards, and Measurement at the National Bureau of Standards* (Boulder CO: National Bureau of Standards, 1986).

[10] Between the time of the first meeting of Hazeltine and Wheeler and the time of this agreement, Wheeler invented another form of neutralization, known as "capacity-bridge neutralization," for which Wheeler obtained the first of his many US patents. (See *Early Days*, pp. 166–169.)

[11] See Wheeler's *Hazeltine the Professor*, which contains two autobiographical sketches by Hazeltine.

[12] *Proceedings of the IRE*, vol. 6, 1918, pp. 63–98.

[13] W. Rupert Maclaurin, *Invention and Innovation in the Radio Industry* (New York: Macmillan Company, 1949) (reprinted in 1971 by Arno Press, New York), pp. 127–129.

[14] *Hazeltine the Professor*, p. 106.

[15] *Hazeltine the Professor*, p. 87.

[16] *Early Days*, p. 117.

[17] G. Breit and M. A. Tuve, "A radio method of estimating the height of the conducting layer," *Nature*, vol. 116, 1925, p. 357. A more complete report is Breit and Tuve's "A test of the existence of the conducting layer," *Physical Review*, vol. 28, 1926, pp. 554–575.

[18] H. A. Wheeler and F. D. Murnaghan, "The theory of wave filters containing a finite number of sections," *Philosophical Magazine*, vol. 6, 1928, pp. 146–174.

[19] Interview 1991, pp. 35–36.

[20] "Automatic volume control for radio receiving sets," *Proceedings of the IRE*, vol. 16, 1928, pp. 30–39.

[21] "Classic circuits," *Electronics*, vol. 53, no. 9, 1980, pp. 436–442. Another

section of the same issue of *Electronics* was entitled "Great innovators," and one of the people thus honored was Harold Wheeler.

[22] Interview 1991, pp. 4–6.

[23] Interview 1991, p. 11.

[24] Interview 1991, p. 6.

[25] Wheeler's work was first publicly presented in a paper given in 1934 by Hazeltine engineers Daniel Harnett and Nelson Case, "The design and testing of multirange receivers," *Proceedings of the IRE*, vol. 23, 1935, pp. 578–593. Wheeler gives a full account of the history and his role in it in Wheeler Monograph Number 8, "The piston attenuator in a waveguide below cutoff" (Great Neck NY: Wheeler Laboratories, 1949). As this monograph points out, independently of Wheeler's work and slightly earlier, John Dreyer, an engineer working for Atwater Kent, invented a form of the piston attenuator, but he was unable to calculate the rate of attenuation, so he had to rely on measurements.

[26] H. A. Wheeler and V. E. Whitman, "Acoustic testing of high fidelity receivers," *Proceedings of the IRE*, vol. 23, 1935, pp. 610–617; and H. A. Wheeler, "RF inductance meter," *Electronics*, vol. 20, no. 9, 1947, pp. 105–107.

[27] "Microwave measurements," *Proceedings of the IRE*, vol. 50, 1962, pp. 1207–1214.

[28] Interview 1991, p. 8.

[29] See *Early Days*, pp. 229–231.

[30] H. A. Wheeler and W. A. MacDonald, "Theory and operation of tuned radio-frequency coupling systems," *Proceedings of the IRE*, vol. 19, 1931, pp. 738–805.

[31] Most important were "Two-signal cross-modulation in a frequency-modulation receiver," *Proceedings of the IRE*, vol. 28, 1940, pp. 537–540, and "Common-channel interference between two frequency-modulated signals," *Proceedings of the IRE*, vol. 30, 1942, pp. 34–50. When Wheeler was presented the Armstrong Medal of the Radio Club of America in 1964, the award citation read, in part, ". . . it is particularly appropriate that this year which marks the 25th Anniversary of FM broadcasting, the Armstrong Medal is awarded to one who also pioneered in the field of frequency modulation. His theoretical analysis of frequency modulated signals helped outline the boundaries of this new discipline."

[32] In a history of radio, Gordon Bussey comments that the rate of progress in radio design in the late 1920s in the United States was slowed by "time-wasting legal arguments over patents" (*Wireless: The Crucial Decade* [London: Peter Peregrinus, 1990, p. 49].) An example of how time-wasting the legal actions could be is provided by Edwin Armstrong's 1949 suit against RCA and NBC for infringement of his FM patents, for which Armstrong spent, in sum, one year on the witness chair (John D. Ryder and Donald G. Fink, *Engineers & Electrons: A Century of Electrical Progress* [New York: IEEE Press, 1984, p. 82].)

[33] The story is much more complicated than is here suggested. Two patents were actually involved: the original diode AVC patent (US Patent 1879863) and a reissued patent (Reissue 19744). In response to several unfavorable court decisions concerning the original patent, Wheeler filed a reissue

application that made more specific claims. The cases referred to in the text concerned the reissue. (A fuller account of the litigation is contained in Wheeler's *Early Days*, pp. 232–248.)

[34] Quoted on page 182 of W. Rupert Maclaurin's *Invention and Innovation in the Radio Industry*.

[35] Floyd L. Vaughan, *The United States Patent System* (Norman OK: University of Oklahoma Press, 1956) (reprinted by Greenwood Press, Westport CT; 1972), p. 22.

[36] Quoted on page 10 of Steven Lubar's "New, useful, and nonobvious," *American Heritage of Invention and Technology*, vol. 6, no. 1, 1990.

[37] As mentioned above, the AVC circuit was honored by the journal *Electronics* as one of twelve classic circuits. Frederick Terman, in *Radio Engineers' Handbook* (New York: McGraw-Hill, 1943, p. 639) credits Wheeler with the invention of diode AVC. The perception that the courts were not protecting patent rights and a 1939 speech by President Franklin Roosevelt condemning further inventions (Roosevelt said they reduced employment) prompted the National Association of Manufacturers to argue publicly for the value of inventions and of patent rights. In 1940 the NAM honored about 100 inventors with the "Modern Pioneer Award"; Wheeler was one of those honored.

[38] Additional information about the two cases settled by the Supreme Court is contained in the interview conducted by the author and in an 8-page manuscript, headed "U.S. Supreme Court," that Wheeler wrote on 27 July 1991, a copy of which is available at the Center for the History of Electrical Engineering.

[39] Albert Abramson, *The History of Television, 1880 to 1941* (Jefferson NC: McFarland & Company, 1987).

[40] H. A. Wheeler and A. V. Loughren, "The fine structure of television images," *Proceedings of the IRE*, vol. 26, 1938, pp. 540–575.

[41] H. A. Wheeler, "The interpretation of amplitude and phase distortion in terms of paired echoes," *Proceedings of the IRE*, vol. 27, 1939, pp. 359–385.

[42] H. A. Wheeler, "Wide-band amplifiers for television," *Proceedings of the IRE*, vol. 27, 1939, pp. 1078–1086. This paper was reprinted for its historical importance in the August 1984 issue of *Proceedings of the IEEE*.

[43] Interview 1991, p. 64.

[44] See Leslie Curtis, "Detectors for buried metallic bodies," *Proceedings of the National Electronics Conference*, vol. 2, 1946, pp. 339–351.

[45] A different system, the Mark IV, was developed in the United States, but a Combined Communications Board decided to adapt the British Mark III for use by all the Allies. See Henry Guerlac's *Radar in World War II* (Tomash Publishers and American Institute of Physics, 1987), pp. 367–374.

[46] Interview 1991, p. 10. In 1936 he published "The design of doublet antenna systems" (*Proceedings of the IRE*, vol. 24, 1936, pp. 1257–1275), and in 1937 he applied for a patent on a horizontal figure-eight antenna.

[47] Interview 1991, pp. 103–104.

[48] M. D. Fagen, editor, *A History of Engineering and Science in the Bell System: National Service in War and Peace (1925-1975)* (Bell Telephone Laboratories; 1978), p. 24.

[49] Ibid., pp. 370–383.

[50] Interview 1991, p. 54. The antenna is described on page 389 and pictured in its radome on page 390 of Fagen, *A History of Engineering and Science*.

[51] Interview 1991, p. 74.

[52] Especially important are the following papers: "Fundamental limitations of small antennas," *Proceedings of the IRE*, vol. 35, 1947, pp. 1479–1484; "Small antennas," *IEEE Transactions*, vol. AP-23, 1975, pp. 462–469; "Antenna topics in my experience," *IEEE Transactions*, vol. AP-33, 1985, pp. 144–151.

[53] Richard C. Johnson and Henry Jasik, *Antenna Engineering Handbook*, 3rd edition (New York: McGraw-Hill, 1993), Chapter 6.

[54] Interview 1991, p. 51.

[55] *Proceedings of the IRE*, vol. 36, 1948, pp. 1392–1397.

[56] The article is in *IEEE Transactions*, vol. MTT-13, 1965, pp. 172–185; the designation of it as a Citation Classic is in *Current Contents*, 2 June 1980, p. 16.

[57] Interview 1991, pp. 89–90.

[58] In *Hazeltine the Professor,* p. 94.

[59] Interview 1991, p. 40.

[60] National Bureau of Standards, *Circular 74* (Radio instruments and measurements), first edition 1918, second edition 1924. In 1932 Frederick Terman wrote (in *Radio Engineering*, p. 669), "This book is the standard authority on the subject [of calculating inductance, mutual inductance, and capacity] and contains formulas for making calculations of any desired accuracy for almost every case than can be encountered in practice." For information about *Circular 74*, see pages 52 through 55 of Wilbert Snyder and Charles Bragaw's *Achievement in Radio*.

[61] See the author index (pp. 997–1004) of *Radio Engineers' Handbook* (New York: McGraw-Hill, 1943).

[62] Interview 1991, p. 33.

[63] H. A. Wheeler, "The real economy in engineering," *Proceedings of the IRE*, vol. 34, 1946, p. 526.

[64] The quotation is from a 17-page manuscript, written by Wheeler, entitled "Selected papers by Harold Alden Wheeler," a copy of which is available at the Center for the History of Electrical Engineering.

[65] "Simple inductance formulas for radio coils," *Proceedings of the IRE*, vol. 16, 1928, pp. 1398–1400.

[66] "A simple formula for the capacitance of a disc on dielectric on a plane," *IEEE transactions*, vol. MTT-30, 1982, pp. 2050–2054; "Inductance formulas for circular and square coils," *Proceedings of the IEEE*, vol. 70, 1982, pp. 1449–1450.

[67] *Proceedings of the IRE*, vol. 30, 1942, pp. 412–422.

[68] H. A. Wheeler, "Transmission-line properties of a strip on a dielectric sheet on a plane," *IEEE Transactions on Microwave Theory and Techniques*, vol. MTT-25, 1977, pp. 631–647.

[69] *Proceedings of the IRE*, vol. 34, 1946, p. 526.

[70] Wheeler's chairmanship of the Technical Committee on Radio Receivers, which lasted from 1932 to 1938, ended with the issuance of a report that soon became the standard guide, used in many countries, for testing radios.

[71] *Early Days*, pp. 16, 20.

[72] Wheeler argued as follows: The CGS system has a set of electromagnetic units and a separate set of electrostatic units, so the practitioner often had to convert from one set to the other; the MKS system had the additional advantage that wavelengths of electromagnetic radiation were being measured in meters (Interview 1991, p. 23).

[73] See H. A. Wheeler, "A logical date code for communications and records," *Journal of Industrial Engineering*, vol. 18, no. 4, 1968, pp. ix–x. Wheeler considers his form preferable to the usual European form of day-month-year because it is already the established practice to put the units in decreasing order when expressing magnitudes (as a number of three digits gives hundreds, tens, and ones).

[74] See *Early Days*, pp. 9–10.

[75] Interview 1991, p. 191.

[76] Manuscript autobiography prepared for Project LMS (1990), p. 8.

CHAPTER 6

Calculating Power
Edwin L. Harder
and Analog Computing in
the Electric Power Industry

EDWIN L. HARDER

Figure 1. Edwin L. Harder, power engineer, inventor, builder of the Anacom computer, former president of the American Federation of Information Processing Societies, and current resident of Pittsburgh, Pennsylvania.

In today's marketplace, laptop computers can move from a market leadership position to obsolescence within a few months. In even the most staid sectors of the computer industry, manufacturers replace their product lines with more competitive products every several years, and one has to look hard to find a company using computers built before 1980. A leading historian of computing has called the first digital computers, built in the late 1940s and 1950s, "dinosaurs"—a description that accurately conjures up images of enormous masses, too slow and cumbersome to survive in a modern environment.[1] It is truly remarkable, then, that in 1990 a general-purpose computer built in 1948 was still in operation at a leading engineering firm in the United States.[2]

This paper describes this computer, the Anacom, and the man who built it and managed its operation for two decades, Edwin L. Harder.[3] Although Harder characterized the Anacom as his top achievement, in the twenty years prior to building this machine he had already distinguished himself as an engineer in the electric power industry. This career led him directly into a leadership position in the computer field. Harder was the author of numerous inventions in his early career, and several of his patents helped to keep his employer, Westinghouse Electric Company, solvent in the depths of the Great Depression. His influence extended beyond his company to the wider professional community through his distinguished service in national and international professional organizations.

Early Life and Education

Edwin L. Harder was born on April 28, 1905, in Buffalo, New York, to Edwin P. and Cordelia Mandana Cousins Harder, the second of five children. His father, having been one of the first to receive an electrical engineering degree from Pennsylvania State University, was superintendent of the distribution substations of the Cataract Power and Conduit Company in Buffalo, which later was consolidated into the Buffalo General Electric Company.[4] As the only college-educated employee, he was on constant call to preside over problems with the Buffalo power network. Harder's mother held various professional jobs, including nurse and teacher, and once worked as a secretary at General Electric. She met her husband while nursing him to recovery from burns received in a manhole fire.

Harder had an active, middle-class childhood. As an adolescent, he worked weekends in the local fruit market. His hobbies included bicycling, roller and ice skating, and raising rabbits and pigeons. He built his own ham radio rig, except for the expensive earphone, which he borrowed from a friendly telephone repairman. His high school, Lafayette, was the best scholastic preparatory school in Buffalo, and he excelled in his studies there. He showed an interest in his father's profession at an early age, and as a high school junior he worked one summer in the meter department at Buffalo General Electric to gain first-hand experience of electrical engineering.

Harder matriculated at Cornell University in the fall of 1922. His father wanted him to attend Penn State, but a four-year, land-grant scholarship to Cornell decided the matter. After giving serious consideration to mechanical engineering, Harder majored in electrical engineering.[5] Cornell had one of the top programs in the country. As preparation to work in industry, engineering students received hands-on training in foundry, forging, woodworking, machine shop, and mechanical drawing. There were three years of physics (including electricity) and chemistry instruction, and an unusually heavy dose of mathematics including differential equations and Heaviside's operational calculus. Electrical engineering courses began with DC circuits and resistors and proceeded to AC circuits. Extensive laboratory work was part of both mechanical and electrical engineering study. Harder took particular inspiration from the teachings of Vladimir Karapetoff.

Although Harder remembers devoting many hours to chess playing and other idle pursuits during college, he compiled an excellent academic record. He was one of a very few students elected to all three honor societies: Phi Kappa Phi, Eta Kappa Nu, and Tau Beta Pi. When he graduated in 1926, the economy was booming, jobs were abundant, and he received several job offers. He selected Westinghouse, where several of his friends from Cornell had accepted jobs and where he believed he would have the greatest opportunity to choose his line of work because of the large number of engineers the company hired each year.

Electric Utility Engineer

Harder was one of about 300 new engineers hired by Westinghouse that year. All new engineering hires were subjected to a year of in-house training. About 100 of these trainees went into engineering, the rest into sales. All of them spent a month or two in each of several departments: building switchboards and wiring circuits in the shop, doing elementary tasks in the design of motors in the motor engineering department, testing big machines, and working in the department that made electrical insulators. For those entering into engineering, the training also included a rigorous three-month engineering school, in which the principal engineers from different departments gave lectures on practical and theoretical subjects. The courses were rigorous enough that the University of Pittsburgh gave graduate credit for them. By taking several additional courses, Harder was able to earn a master's degree in electrical engineering there in 1931.

I have an interesting story about my Master's thesis. Pitt has a catalog out that tells the requirements of getting a master's degree or a doctor's degree. And it says plainly that the thesis has to be published. So I get this problem of DC in transformers. First time it's ever happened. Beautiful thesis project! All this was brand new information. Nobody knew anything about it before then, and so I made tests, found out how it worked, and wrote it up and published it. Took it down to Pitt and— "Oh, no! It doesn't mean that. We have to approve the thesis before you start working on it. You can't use that for a thesis." "Why?" I said, "the book says it has to be published." "Well, it doesn't mean that. We have to approve it first." So then I came on this saturating reactor problem, and I did get it approved ...that I could write the thesis on this subject. I happened to be down in Professor Dyche's office one day, and I said, "Well, it's coming along fine. It's going to be published in the Electric Journal. *It's going to be published in March." I was probably down there in November or December. Dyche almost blew up. He said, "Well, you can't publish that." I said, "The book says it has to be published." He said, "Well, it doesn't mean that. What it means is that* **we** *publish it." And so we called Charlie Scarlott, who was the editor of the* Electric Journal, *and Charlie*

agreed to delay it for several months before he published it so that I could add a
footnote that this is based on a thesis presented at the University of Pittsburgh. So
he published it, after the thesis was accepted. He was a good friend, and he
delayed it. So that's the story of how this finally became my master's thesis. But
it also was a good subject because it was something that had never been run into
before and had a lot of good theory connected with it.[6]

Harder's mathematical training at Cornell helped to set him apart from
the other new recruits. In one of the in-house courses, he caught the
attention of an instructor by being able to derive mathematically the
equations of traveling waves on transmission lines. This helped him to
secure a position in the General Engineering Department, which worked
on systems problems, mainly consulting for customers with difficult
technical problems that occurred in the application of electrical equipment
to their systems.[7] All General Engineering employees began by working for
two years in one of the design engineering departments in order to get to
know a particular kind of equipment manufactured by the company.
Harder was assigned to the electrical development section of the Power
Engineering Department, which built large rotating equipment.

Although Power Engineering was ostensibly a good place to learn about
the company's large rotating machines, Harder's section head was totally
immersed in a problem relating to air flow and ventilation of machines, and
Harder was assigned nothing but these kinds of problems to work on. His
experience here was no doubt narrower than the company policy intended,
but it taught him about dimensional analysis in models, which he later put
to good use in his computing work. While working in Power Engineering,
he made his first significant contribution to the company by constructing
a hot-sphere anemometer, a device conceived by one of his fellow engineers,
which was inserted through the bolt hole of a machine to measure the
internal air velocity.

By accident, Harder was able to return to General Engineering before his two
years were up. Power Engineering laid him off during a company downsizing,
but just at that time a position became available when someone left General
Engineering after it had already made its required personnel cuts. Harder was
assigned to railroad electrification—a fortuitous assignment because it enabled
him to remain employed during the Depression. The Pennsylvania Railroad
undertook a major program to electrify its lines from Philadelphia to West
Chester, Trenton, New Brunswick, New York City, Washington, and Harris-
burg during the Depression. In the darkest days of the 1930s, the railroad's
president, General Atterbury, ordered ten million dollars worth of electric
locomotives from Westinghouse. This order helped to keep the East Pittsburgh
plant open and Harder employed.[8]

Harder was attracted to General Engineering because it concentrated
on systems work rather than on the design of specific equipment. The

prevailing characteristic of his work there, however, was extensive calculation, including the use of (digital) desk calculators and various special-purpose analog devices built by the company. In college Harder had never used a calculating tool other than a slide rule. But in General Engineering he soon became responsible for using an electrically powered Marchant desk calculator to make extensive calculations on the engineering design for the electrification of the Pennsylvania Railroad.[9] Another project, which investigated rectifiers used in the power transmission system that were causing telephone interference, required him to use a Chubb mechanical harmonic analyzer, a device similar to a planimeter that was moved mechanically around a cutout of a polar oscillogram to determine the rectifiers' harmonics.[10] Based upon these calculations, he designed AC and DC filters to reduce the harmonics to a level at which they no longer caused serious interference.[11]

From 1938 until 1946, I was assigned to the Middle Atlantic District, including Washington, D.C., as a sponsor engineer. In the latter part of that period, we were visiting the Potomac Edison Company, which is headquartered in Hagerstown, Maryland. It's a system about 100 miles wide, maybe from Cumberland to Fredericksburg [Maryland], about 50 miles north-south from the Pennsylvania Turnpike down into West Virginia. Their headquarters was in Hagerstown, and the salesmen from Philadelphia would take me down there occasionally. They had one good technical man, and they had an older manager who had previously run the streetcar system from Pittsburgh to Butler [Pennsylvania]. He did a commendable job, but they were having a very unusual problem; at nights and [during other] light-load periods the whole outskirts of their system would have a very high fifth-harmonic voltage content. It would get 10 or 15 percent fifth-harmonic voltage. On a capacitor, 100 percent of 60-cycle voltage will produce 100 percent current, but it only takes 20 percent of fifth harmonic to produce 100 percent current because the impedance of the capacitor is only one fifth as much. The older capacitors did not have enough reserve in them to handle that much extra current, and consequently they were failing. Also, switchboards were burning up due to excessive currents in switches.

They took me up to the Pennsylvania Turnpike and showed me some of these stations where the switches were all blistered. They were very confused. As I told you before, I had had a lot of experience with harmonics. I knew harmonics better probably than anybody because I had had this dual experience of telephone interference and building filters and teaching harmonics, and the Pennsylvania Railroad's problem with harmonics. They didn't have any rectifiers and the only source of harmonics is the transformers. If a transformer has to produce a sine-wave voltage, it has to have a sine-wave of flux. And the exciting current is not going to be sine-wave. It's going to have a lot of harmonics in it. I knew a couple of different equivalent circuits of transformers that correctly represented them. I knew that up at Sharon we had used several kinds of iron in transformers. Just from the characteristic of the iron, I could calculate what these harmonics would be. It wasn't too different for any of these five different kinds of iron they used. I assumed that GE's iron must be about the same and that this equivalent circuit would probably work for all transformers.

So I set up the whole Potomac Edison system on the AC calculating board at 300 cycles (i.e., at fifth harmonic). I started with the equivalent circuits of the transformers which is where the harmonics were coming from. I represented all the loads, mostly guessing at them. The starting current of a motor might be five times normal. Well, that's when the slip is 100 percent, when the motor is standing still. The fifth harmonic is the same. So motor load is roughly 20 percent impedance at the fifth harmonic. I guessed at how much was lighting and static loads, and how much was motors, and I put the loads at all of these stations. Then I ran it on the calculating board, which showed that they were right at the peak of a resonance. If you added capacitors to the system you passed the resonance point.

So I told them that if they would install a 2,000 kVA capacitor at Winchester, Virginia, it would solve all their problems. Nobody had ever done this before or has done it since, so far as I know. A 2,000 kVA capacitor was at least ten times bigger than anything they had ever installed before in the way of a capacitor. They had installed 100 or maybe 200, but not 2,000 kVA. So they asked GE about it, and GE wrote them a letter stating that it wouldn't work and recommending against it.

But the technical engineer there was a radio ham, too, and knew a lot about electricity. He knew enough that he could go to a station and filter out the fifth harmonic and put it in an oscillograph and measure it. He could get the right circuits set up so that with an ordinary oscillograph he could take a picture of what the fifth harmonic was. He saw my calculations and asked why I concluded this. He believed my reasoning, and they installed the 2,000 kVA capacitor at Winchester. There were tests, and it did exactly what I said it would do.

I had told them if they put a coil—a reactor—in series with this capacitor and tune it to the fifth harmonic, the current that it will draw will be within its capacity. I had measured on the calculating board how much this current would be, and there wasn't too much. They could actually tune it, so they called it a filter. They put this capacitor on, and they got the benefit of it for power factor correction. It actually short-circuited the fifth harmonic at Winchester. And at Frederick, 50 miles away, it brought it down to half. There was a fellow from the telephone company there to witness the test, and he wrote me after he retired he had been so impressed by that test; nobody had ever tried it before. I wouldn't have tried it if I hadn't had all that experience with harmonics. And I had confidence in my calculations. [12]

During this period Harder also did some important calculations for the company on the effect of lightning strokes on transmission lines. One of

Westinghouse's chief engineers, C. L. Fortescue, had determined that only direct lightning strikes cause damage to transmission lines, not strikes in the vicinity of the lines as had been previously supposed. Working with his mentor in the company, A. C. Monteith, Harder made calculations on a design to make transmission lines safe from lightning strikes. Knowing how many direct strikes occurred on average on a mile of line per year, and knowing their average current, Harder calculated the grounding resistance of the towers so that the current would be carried through the ground wire and ground at the tower, without creating a high enough voltage to flash over the line.[13]

Because of his experience with these calculations, Harder helped to set the design specifications, such as the sizes of the resistors, for a special-purpose AC Calculating Board the company built in 1929.[14] Although it could solve only fixed-frequency problems, it superseded the numerical methods and desk calculators Westinghouse had previously used for these purposes.[15] These boards were built in quantity by Westinghouse and sold to electric utility customers all over the world. The AC Calculating Board continued in operation well into the 1960s, until it was finally replaced by digital computers.

In 1933 Westinghouse shut down the General Engineering Department as a Depression-era economy measure, and Harder transferred to the Switchgear Department, where he remained for five years (except for a six-month transfer in 1936 to the Relay Department in Newark). His main work during this time was to design the relay system for the electrification of the Pennsylvania Railroad.[16] He also worked on the designs of the switchboards for the Hoover Dam and of voltage regulators for the Safe Harbor Water Power Plant on the Susquehanna River.[17]

In 1938 Monteith, who had become Manager of Central Station Engineering, hired Harder as his engineering representative for the Middle Atlantic District, which included Washington, D.C. Harder remained in this position through the war. During the war, he worked with Captain Hyman Rickover to develop the power supply systems for the new naval vessels that were being built to replace nineteen American warships the Japanese destroyed at Pearl Harbor. These power systems did not drive the ships, but they provided all the on-board power. The challenge was to make power supplies that could swing heavy gun turrets and raise elevators, without dropping the voltage so much that it would disable the operation of the ship's radar and other sensitive electronic gear.

General Electric and Westinghouse collaborated on this project, each assembling its best engineers to work together on generators, exciters, turbines, governors, and voltage regulators. Harder handled the system work for Westinghouse, while Sil Crary did the same job for General Electric. For this purpose, Harder invented and built an electronic device, known as a regulating system simulator, to calculate voltage drop from sudden loads in systems that incorporated turbines, generators, governors, and voltage regulators. The simulator saved many hours of long-hand calculation, and Harder later used the simulator as a component in the Anacom.

Harder's war work was the basis for his doctoral dissertation, begun at the University of Pittsburgh in 1945. His dissertation gave a general solution to the voltage regulator problem.[18] Recognizing that there were just a few variables in a regulator system, Harder varied them all in turn and was able to determine from these trials the basic operating properties of a regulator system. His main result was that two-delay systems (e.g., ones involving an excitor and a generator) are always stable, whereas systems involving more delays are not necessarily stable, but may be made stable for a particular set of parameters through the use of feedback.

After World War II, Westinghouse had problems with many of its regulation systems for steel mills, paper mills, and generators, which cost the company millions of dollars. The company might get a steel mill in Gary running properly, but when the mill wanted to roll thicker or thinner steel, or to speed up or slow down the milling line, the process would frequently become unstable and the engineers at Westinghouse would be called back to restabilize the system for the new operating parameters. The Westinghouse engineers did not at first recognize any general principles in the problems they were experiencing, and they attempted all sorts of ad hoc solutions. Harder's dissertation provided a general answer to the problem, although even Harder did not fully understand the general situation until many years later. These mills were three- or four-delay systems, which could be stabilized for certain parameters, but which would become unstable when the parameters changed. Once magnetic amplifiers became readily available, Westinghouse used them to build what were essentially two-delay regulating systems for these mills, thereby providing stable operating environments despite variations in the operating parameters.[19]

Harder's career took a new direction in 1946, when one of Westinghouse's most impressive young engineers, G. D. McCann, decided to return to his alma mater, California Institute of Technology, as a faculty member in electrical engineering. Harder was selected to replace McCann as the company's consulting transmission engineer, director of lightning studies, and director of the recently conceived Anacom computer. In these positions, Harder gave many lectures and consulted widely on behalf of the company. He had considerable independence and worked on a number of interesting and challenging projects. One project was a study to measure the current in lightning strokes by placing recording equipment on top of the Cathedral of Learning at the University of Pittsburgh and on fire towers around Maryland, Pennsylvania, and West Virginia.[20] Among these duties, the most important turned out to be the one to build and then direct the use of the Anacom. At first, he had only one or two people working for him, but eventually his staff included 140 professionals in a unit that came to be known as the Advanced Systems Engineering and Analytic Department.[21]

Harder as Inventor

Especially during his first twenty years at Westinghouse, through the end of World War II, Harder was a fecund inventor. His Westinghouse career resulted in 123 patent applications and 66 patents. After the war, he assumed extensive managerial responsibilities, and his patent activity subsided. He regretted the lack of time for research.

As a young engineer, Harder was encouraged to return to Westinghouse at night and work on patentable ideas. He apparently did this with alacrity, happy to have the laboratory facilities entirely at his disposal. He felt an obligation to the company and to society to apply himself to invention:

> I had that [inventive] ability, and I wasn't making the most of it. I was turning in some patents, but in almost any field that I turned to I could see better ways of doing what they were doing. If I had that ability and I only worked eight hours a day at it and goofed off the rest of the time, I really wasn't making full use of my potential. It certainly made life interesting to have these problems coming up that seemed very difficult to solve and then come up with a unique solution to it that nobody else had thought about. It would occur to you, and that's a great feeling. That's almost like winning a football game.[22]

There was a pattern to Harder's inventive process, as he explained: "Many of my inventions were highly technical, or mathematical, in nature. The problem would be expressed mathematically, and I would try to find apparatus that would do what the resulting equations said had to be done. Some of my finest and most valuable inventions were a result of this process."[23] Harder was unusual in the power engineering field for the degree to which he relied on mathematics. His doctoral dissertation on voltage regulation was awarded by the Pitt mathematics department, not by electrical engineering. His last twenty years as a regular employee at Westinghouse were spent in the Analytical Department, where he mainly applied mathematical methods and computing machinery to problems of power engineering.

Harder's patents include several on analog computing, but most of them cover aspects of power engineering—mainly control of power systems, including numerous inventions of regulating and relay systems. Harder regards four inventions as his most important: the HCB relay, a high-speed distance-type carrier-pilot relay system, the economic dispatch computer, and a linear coupler. His other inventions, he argues, were incremental improvements to large fields.

The first of these inventions was the HCB (high-speed, current balance) relay, which was a new design for tripping the circuit breakers of a faulty transmission line.[24] The transmission lines of a power system generally form an interconnected network. When a fault occurs on any one of these lines, it must be disconnected from the system without disturbing the other lines in order for the power system to continue to function safely and properly. The highest speed protection is obtained through the use of a pilot circuit, essentially a telephone line over which a relay is run in order to compare the currents at the two ends of the line. If they are the same, then there is no fault on the line. If they are different, there must be a fault on that line section and it needs to be tripped out. For long transmission lines, instead of stringing long lengths of extra wires for the pilot circuit, a 60-cycle current is carried on the transmission line itself for this same function (known as carrier current pilot circuit). For short lines, pilot wires between the two ends are most economic, either carried in a control cable along the line or rented from the telephone company. The HCB is a type of pilot wire relay.

Ten kinds of faults can occur on a transmission line between the phase wires, a, b, and c, and the ground g: wire-to-wire (ab, bc, ca, abc), wire-to-ground (ag, bg, cg), multiple-wire-to-ground (abg, bcg, cag). It would be desirable to have a single voltage that could be compared over the pilot wires, regardless of which of the ten kinds of faults occurred; and this voltage must have a respectable value (i.e., one large enough to be read at both ends of the line), whichever kind of fault occurs. Through the method of symmetrical components, a mathematical technique well known to power engineers, Harder determined a network of resistors and reactors that would produce such a "discriminating" voltage when the a, b, and c currents were passed through it. This was the HCB relay. A simple operating element closes if the discriminating voltage differs at the two ends of the line and stays open otherwise. The network of resistors and reactors was inexpensive relative to the existing carrier relays, which comprised nine or more operating elements.

Harder invented the HCB relay in 1932, while he was in General Engineering, but it was not put into production at that time because of the Depression. In 1937, when it finally did become a company product, it was unknown to the Westinghouse relay department in Newark, which normally would have had responsibility for its development. About 1937, Public Service of Indiana was building a loop of short transmission lines around the city of Terre Haute, which were to be protected by pilot wire relays. General Electric offered their carrier current scheme, simply substituting a DC pilot wire signal for the carrier signal, thereby providing one-cycle (1/60 second) relaying. Customers were much more concerned about response time than they were about cost. They wanted to clear faults as quickly as possible in order to minimize damage and resume full operation, and Westinghouse had no comparable system with one-cycle response to offer in competition to GE's. Here is where the HCB story begins, as told by Harder.

The salesman from Indianapolis, Freddie Green, came into Switchgear one day, and he was crying on my shoulder about Westinghouse having nothing to offer for this job and General Electric getting the order without any competition. And I said, "Well, I have this system that I patented four or five years ago. One weekend, a few months ago, when I was testing some of the carrier current relays on our miniature transmission line in East Pittsburgh, I took the weekend and went in and scrounged enough resistors and reactors to make up two of the required networks and tested it. It tripped for all the internal faults, and it didn't trip for any external faults." Now, I said, "It's completely undeveloped. If you want to take a chance on it, it's up to you." Well, he said, "What have we got to lose?" So we got Newark to put a price on it, and they priced it about half of what the General Electric one was. We still had a tremendous ratio of sales price to cost because the relay had only one little element that opened and closed. All the rest of the brain was in the network. These resistors and reactors, which were very inexpensive, had nothing critical at all about them. So it boiled down to a meeting with Joe Trainer, who was the chief engineer of the Public Service Company of Indiana, which ran the Terre Haute operation, and Arbuckle, the relay engineer, who had practically promised it to GE. He had done this since there was no competition. Now he had to go in to his boss and explain. He said, "Here Westinghouse is offering this relay that only has one contact. If that fails we don't have any protection." Joe Trainer looked him right in the eye and said, "All these years you've been wishing that instead of a whole mess of contacts you could have one contact that closed when it should and doesn't close when it shouldn't. And now that they're offering you exactly that, you don't want it. I don't understand." So the result was that GE didn't get to use their GMB [carrier current relay] even for the pilot-wire application.

That HCB relay is still widely in use [several hundred are sold each year, sixty years after invention]. Patents only last 17 years so obviously anybody could build it, and [certainly] in the last few years [they have]. But there's a tendency for companies not to just pick up another company's product and start building it. So it's still largely a Westinghouse product. And it was worth many millions of dollars to the company in business.[25]

Harder's second major patent was for a high-speed, distance-type, carrier-pilot relay system. He developed this idea in 1936 during a 6-month assignment to the Westinghouse Relay Division in Newark, New Jersey.[26] During World War I, the Army Corps of Engineers began using carrier current (i.e., a few watts of 50 to 150 kHz current) to transmit messages over long power-transmission lines. After the war, power companies used this method for protective relaying circuits and communication over transmission lines. By 1930 both Westinghouse and General Electric were offering carrier current relaying. When a fault occurred, the fault detectors at each end of the line would start transmitting carrier. If the directional relay at terminal A pointed into the line, it would stop carrier transmission from that end. If the directional relay at terminal B also pointed into the line, then the fault must be on the line section between A and B. It would also stop carrier, and both ends would trip simultaneously, with no delay.[27]

The best previous relay scheme used distance-measuring relays, but involved sequential tripping for faults near either end of the line, a delay of about a half second for the second breaker. Until 1936 both General Electric and Westinghouse were using conventional three-phase directional elements, operating in three to five cycles. The circuit breakers then took about eight cycles to open. However, unbeknownst to Westinghouse, General Electric had spent a full year developing a very fast three-phase directional element and was now offering a one-cycle carrier scheme. It was almost impossible to sell a 3- to 5-cycle system in competition with a 1-cycle system. This jeopardized the business of not only the Relay Department in Newark but also the Switchboard Department in East Pittsburgh because about half of the relays were sold with switchboards. Many new transmission lines were being built at this time, and Westinghouse stood to lose many millions of dollars worth of relay and switchboard business. As Harder remembered the situation, "Not only were we a year behind, but how in the world were we going to develop a high-speed, three-phase directional relay without infringing on their [GE's] patents? We just had to have some interim scheme to offer in the meantime—no matter how crazy."[28]

Harder devised a scheme using the high-speed, single-phase directional elements of the Westinghouse distance relays, interconnected with many elements for detecting faults. On paper it would be one cycle. But eight separate relay elements, four directional and four fault detectors, all had to open or close correctly, coordinate with each other, and do it in less than 1/60th of a second. The probability of this seemed very remote to the Westinghouse engineers, but none of them had a better idea so they decided to test Harder's scheme. Within a few days, the laboratory engineer reported that the elements did seem to coordinate. After a much larger number of tests were conducted, not a single misoperation had occurred. The company immediately started to sell Harder's system. As Harder recalls:

> We really had the luck of the Irish. . . . It seemed that each job we sold, the customer would find some new advantage of this system that we had not yet thought of. We were just glad to have something that worked at all—in one cycle.
>
> One of the first customers observed that with a three-phase directional element, when you pull the test switch to test the relays, the transmission line has no protection. With the single-phase elements, there are four of them—three phases and ground. Any fault always involves at least two (phase-to-phase or phase-to-ground). Thus the test switch can be pulled on any one of the four to test it and still have 100% protection of the line. When all are in service, there is some redundancy.
>
> Another customer observed that if the carrier was out of service for any reason, the scheme reverted to distance relaying, the very best protection available before the days of carrier.

Still another noted, "Well, I can buy distance relays today and add the carrier later when it is justified."

And so it went. The 1938 paper[29] lists several other advantages of this new carrier-pilot relay system.

The net result was that at the end of the first year we had sold 87 systems and lost only 13 sales. This continued for quite a few years. Westinghouse never did develop a high-speed, three-phase directional element. General Electric abandoned its system after a few years and brought out a system more like ours. Instead of losing millions of dollars worth of business, we gained substantially.[30]

In 1940 Harder made his third major patented invention, a linear coupler, which improved the detection of faults on power buses.[31] A typical high-volume bus on a power system might have several circuits connected to it, perhaps two or three transformers from local generators, and several transmission lines to various parts of the system. There might also be a "bus tie" to another high-voltage bus. Should a fault occur on the bus, it is necessary to trip all of the circuit breakers to it in order to avoid a fire. On the other hand, for a fault on a line just off the bus, you need to trip that line without tripping all the other lines connected to the bus so as to disconnect as little of the system as possible. Theoretically, the sum (totalization) of all currents into the bus will be zero, unless the fault is actually on the bus. Current transformers are typically used in all the circuit breakers to reduce the high primary currents to small secondary currents for this totalizing.

If the bus is at or near a generating station, the short-circuit current contains a large DC transient, which saturates the iron of the various current transformers unequally. The totalized secondary current may be high enough to trip, even if the fault is not on the bus. Many schemes have been devised to avoid this spurious tripping, with varying degrees of success.

Harder's linear coupler omitted the iron and hence avoided the saturation problem.[32] The invention was based on the fact that if the device is a perfect toroid, it is astatic and the totalizing is perfect as long as each circuit goes somewhere through the hole of its toroid. This was the best bus-fault protection available for many years, and even today, half a century later, many of these devices are still in use.

Harder's last major patent, made in 1953, was for an economic dispatch computer.[33] The computer simulated a power generating network comprising generators, transmission network, and loads, in order to optimize the generation of power in interconnected systems, or in the parlance of the electric utility industry, to give it "economic dispatch."

Power companies knew the generating cost for any individual generating unit once the cost of fuel and the efficiency of the unit was known. However, a loss occurred when the power was sent over the transmission lines. If there were only one generating unit, it would not be difficult to determine this loss. A typical network, however, contained perhaps fifteen power stations, each of which affected all the others. To calculate the economic dispatch for a given power station, that is, the amount of power the station should dispatch for the lowest overall cost of the power system, thus required solving fifteen simultaneous equations. This made it difficult to determine the economic dispatch for the system.

Harder convinced one of Westinghouse's customers, Allegheny Power, to have Westinghouse build for them an economic dispatch computer. It was of analog design, with cost curves for individual generating plants represented by resistors, together with summing amplifiers and servomechanisms to solve the simultaneous equations that must be solved to obtain the economic dispatch when considering system losses.[34] The overall transmission loss formula of the system was represented in the computer. The operator could set the generating capacity in megawatts desired from the power network, and the computer would calculate the needed amount of power from each station on the network in order to achieve economic dispatch. (See Figure 2.)

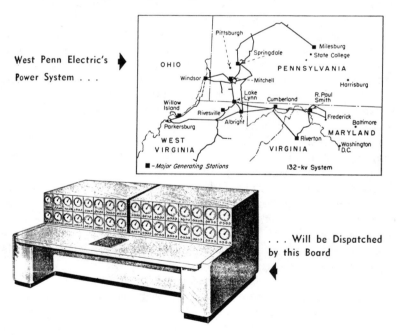

Figure 2.   Harder's economic dispatch computer.

The economic dispatch computer is the one example of an invention in Harder's career upon which he looks back feeling that an important opportunity was missed:

This economic dispatch computer should have been commercialized. No question in my mind that that was a bad mistake. Part of the responsibility is mine. I was young and immature. I should have gone to bat for it, but it never occurred to me that this was an excellent product that should be used. I just did my job getting that [first] one designed, built and installed and then forgot about it. I had other things to do—I was busy. But looking back at it— that was a bad mistake.[35]

Harder's staff built and adjusted the economic dispatch computer and knew how it operated, but the Switchgear Department was responsible for selling. Presumably as a result, no other economic dispatch computers were sold—even though Harder believes that Westinghouse would have been able to sell thirty or forty of them in the national market.

Specialized Calculating Devices at Westinghouse

Harder's greatest accomplishment, the Anacom, was the result of a plan by Westinghouse just after World War II to replace many of the specialized computing devices it had built over the previous quarter century with one, more powerful, general-purpose calculating machine. Before turning to the design, development, and application of the Anacom, we review these early computing devices.

Analog computing instruments and machines were popular among engineers in the early twentieth century for at least four reasons: they were effective at solving problems of interest to engineers; they operated according to principles that engineers readily understood (given that the analog machines were often scaled-down models of the problems under investigation); they did not require engineers to learn abstruse numerical methods, as digital computing devices did; and they could often be built in the engineering laboratory, much like specialized test equipment.

The demand for analog computing devices, especially complex ones, increased through the mathematization of engineering. The expansion of electric power after World War I and a consolidation and increase in the scale and scope of power networks led electrical engineers in particular to develop new analog computing devices. The product integraph and differential analyzers of Vannevar Bush, the network analyzer of Harold Hazen, the simultaneous linear equation solver of J. B. Wilbur, and other analog computing devices constructed at MIT in the interwar years are well known.[36] General Electric, which needed such devices for its work, worked closely with MIT on some of these machines and used them after they were completed.

Westinghouse took a different course from General Electric by building its own machines. The first of these machines was a DC Calculating Board, built by the company after World War I.[37] It was used to calculate short-circuit currents in power systems as part of the process of applying circuit breakers and relays. These calculations became complicated when several power stations were connected to a system, which occurred with increasing frequency after World War I.

Westinghouse engineers determined that a model network containing only resistance could give approximate results to networks also involving reactance and impedance, thus reducing the complexity and cost of the network. The DC network model, known as the DC Calculating Board, which Westinghouse engineers built, consisted of thirty adjustable resistors mounted in three frames, able to take on eleven different resistances between 50 and 1000 ohms. (See Figure 3.) Buses connected the units and subunits. Power was supplied from a 500-watt motor generator able to deliver 50, 100, or 120 volts at up to five amperes. Results were read off an ammeter, which could be connected flexibly to any resistor, and a voltmeter, which could be connected to any point of the network.

Figure 3.   The DC Calculating Board.

Some problems could not be solved on the DC Calculating Board, such as load flow calculations, which involved phase angles and complex impedances.[38] Numerical methods and desk calculators were used instead in these cases.[39] Harder and fellow engineer Frank Green were among the last people to calculate an extensive power system using desk calculators when, in 1928 or 1929, they completed a calculation of the Pennsylvania Railroad transmission/trolley (132 kV/12 kV) system, including the Philadelphia suburban electrification network and the main line extending as far as Trenton, New Jersey.[40] The two engineers worked two weeks at desk calculators to make and check their calculations. The product of these calculations was a table of self- and mutual-impedances to all the load points on this large network.

Desk calculators were replaced around 1930 by an AC network calculator (see Figure 4), and numerical methods were not used in making power calculations again until the mid 1950s, when digital computers such as the IBM 650 and 704 became available.[40] Even after digital computers began to be used for power applications, the AC Calculating Board continued in use for many years because it was easier to set up these power problems on it and because it provided excellent visualization of power system phenomena.[41]

Figure 4. One side of the AC Calculating Board.

The AC Calculating Board had two major advantages over previous calculating equipment used in the electric power industry. One was its ability, unlike the DC Calculating Board, to account for phase-angle differences of voltage in different parts of a network. The other was the incorporation of a whole range of features that made for easy and flexible use: capability to represent a wide range of electrical constants, flexibility in connecting and adjusting circuits, straightforward checking of problem set-up, and simplicity in metering answers.

The AC Calculating Board was used for four general kinds of problems: (1) those associated with operating issues, such as voltage regulation and current and load distribution, of complex transmission and distribution systems involving as many as 36 generating stations; (2) the study of system stability under both steady-state and transient conditions; (3) the investigation of mutual induction between parallel conductors, as in AC railroad electrification; and (4) in short-circuit analysis of interconnected power systems with large power concentrations—for which high accuracy was desirable.

The AC Calculating Board was arranged in a U-shape, each side approximately twenty feet long. The originators, H. A. Travers and W. W. Parker, explained the layout:

> The board consists of fifteen cabinets of standard steel office desk construc-
> tion, in which are located the various units of resistance, reactance, capaci-
> tance, power sources for simulating synchronous apparatus, cord-and-jack
> assemblies to permit reproducing any miniature system network to be
> studied, and suitable current and voltage receptacles whereby the metering
> equipment may be inserted in any component of the miniature network. In
> the middle is located a desk containing the metering equipment [for reading
> phase-angle, voltage, current, and power], as well as ordinary office desks for
> the use of the operators. A synchronous motor-driven generator set supplied
> three-phase, 220-volt, 440-cycle power. Each source of power, representing
> a generator station, consists of a three-phase, 220-volt, 100% induction
> regulator and a phase shifter.[42]

An analysis was carried out by writing an electrical network diagram, scaling the model through the use of an impedance conversion factor, setting up circuits to model the network on the calculating board, adjusting the generating sources, and measuring the results with meters. Individual components were accurate to within 1.5 percent, many errors were compensating, and system solutions were generally within 15 percent accuracy and sometimes much better.

The next Westinghouse special-purpose calculating device was the Electrical Transients Analyzer, built in 1936 according to the design of R. D. Evans and A. C. Monteith.[43] It was used in investigations of various kinds of electrical transients. It included elements of the AC Calculating Board and an 8-stage rotating synchronous switch, which repeated the transient and enabled it to be displayed and measured on a cathode-ray oscillograph. The timing of the opening and closing of the several switch contacts could be controlled, enabling the engineers to investigate complex switching and arcing ground transients. While the AC and DC calculating boards were primarily used for solving short-circuit, steady-state, and stability problems of electric power systems, the Electrical Transients Analyzer and all subsequent special-purpose analog calculating devices built by Westinghouse prior to the Anacom were intended primarily for use on transient problems.

During the war, Harder built a Servo-Analyzer (also known as the

Regulating System Simulator) as part of his work on the power supply systems for the new ships for the US Navy to replace those destroyed at Pearl Harbor. The system enabled him to optimize the voltage regulating systems for these power supply systems. In many respects, the Servo-Analyzer was similar to the Electrical Transients Analyzer. It had a commutator for repeatedly inducing transient voltage or current into the model network and used oscillographs to measure transients. The major addition to this calculating equipment was an electrical analog for a regulating system, consisting of delay circuits (of variable resistance and capacitance), damping transformers, and electronic amplifiers.

In 1945 G. D. McCann, H. E. Criner, and C. E. Warren devised a Mechanical Transients Analyzer. (See Figure 5.)[44] This equipment incorporated the Electrical Transients Analyzer in a system to calculate transient mechanical problems, such as transient torques in the shafts and couplings of turbo-generators during short-circuits. The Electrical Transients Analyzer was coupled with an electrical analog of the mechanical system, in which force, displacement, and velocity in the mechanical system were represented by transient voltages or currents in the electrical analog. Many problems were adaptable to this machine, including mechanical stresses from "plugging" steel mill drives, peak stresses in air frames at the onset of flutter, transient coupling stresses in long railway trains, and transient loading of cranes and hoists. (See Figure 6 for an example.)

Figure 5.   The Mechanical Transients Analyzer.

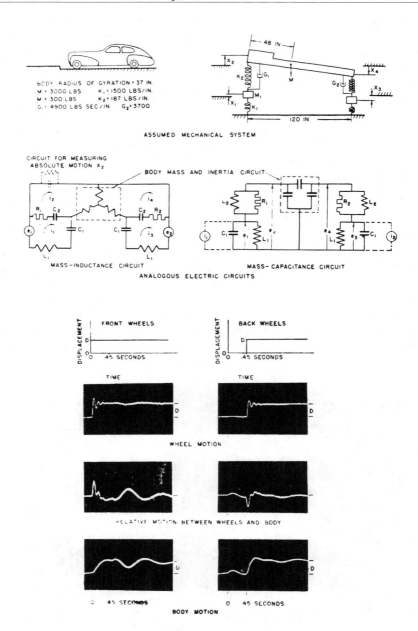

Figure 6. An illustration of the use of the Mechanical Transients Analyzer for studying the ridability of a vehicle. Part (a) shows the assumed mechanical system and the analogous electrical circuits, Part (b) the motion of various parts of the vehicle.

The Anacom

In the 1930s and throughout the war years, electric analog computers were effectively employed in the study of applied mechanics, hydraulics, heat flow, transients in electrical circuits and machinery, and electromechanical systems such as servomechanisms and control.[45] Soon after the war, in 1945, Gordon McCann, a Westinghouse transmission engineer who had made a name for himself within the company for his investigations of lightning, made a survey of all the possible uses of electric analog computers, paying special attention to their potential use in the design and application of electrical equipment. He became involved in this work through his association with R. D. Evans, the codesigner of the electrical transients analyzer, who worked in the same section with him, and through his collaboration with the mechanical engineer H. E. Criner on the mechanical transients analyzer. McCann also made a study of other calculating technologies to determine their suitability to the studies of electrical equipment that were of greatest interest to Westinghouse. McCann deter-mined that a general-purpose electrical analog computer more powerful than any of Westinghouse's existing special-purpose analog calculating devices would be the "best adapted generically and economically to the scope of problems under consideration."[46]

McCann began to design a new general-purpose computer, and proceeded as far as building a small prototype of the Anacom (short for ANAlog COMputer) with his own hands, using the rotating synchronous switch, many resistor and reactor circuits from the AC calculating board, and the servo-analyzer. Harder, who used the prototype for his dissertation research (see Figure 7), was well prepared to replace McCann in the design of the full-scale Anacom, when McCann left for Cal Tech and Harder assumed all of McCann's responsibilities. It was soon arranged for Westinghouse and Cal Tech to cooperate in the construction of similar, general-purpose analog computers at each institution.[47]

Westinghouse allocated $300,000 for the design and construction of the full-scale Anacom. Although it would have been possible simply to cable together a number of pieces of equipment, Harder decided that there was value in building the Anacom to look like a real machine. He used first-class components and had high-quality cabinets specially built to house the various units. He is convinced that this was an important reason why the machine survived for over forty years. The quality construction, together with the unknowns of building such a novel device, however, caused the design and construction costs to mount to $500,000—an overrun which at first upset Westinghouse management, but which was eventually forgiven when the machine began to pay dividends. Part of the expense was in the materials that needed to be used in the full-scale machine. In scaling up, there were serious questions about how to build a large machine without increasing the capacitance of the wiring to a point where it vitiated the results. Special low-loss reactors and transformers composed of unusually large amounts of expensive molypermalloy iron were used to avoid this problem.

Figure 7.   Harder at the small prototype of the Anacom.

Like other analog computers, Anacom had equipment serving three purposes: (1) "forcing functions" to generate electrical voltages representing the forces applied to the physical system under investigation, (2) an analog of the system being studied (i.e., an electrical system governed by the same differential equations that govern the system under investigation), and (3) equipment to measure the changes that occurred when the forcing functions were applied to the analogous circuit. Forcing functions were achieved with 60 hertz and 400 hertz power supplies, an audio oscillator, and a synchronous switch. The analog was built from resistors, low-loss capacitors, hi-Q inductors, and transformers—together with amplifiers when servomechanisms were being modeled. Measuring equipment included ammeters, voltmeters, wattmeters, harmonic analyzers, and cathode ray oscillographs whose readings were recorded photographically.

The choice and arrangement of the units comprising the Anacom were determined by the kinds of problems that were likely to be calculated most frequently.[48] The Anacom occupied a room forty feet long. Along one wall were the circuit element cabinets and the plugboards for connecting them. These were used to form the electrical networks representing the problem analog and the forcing functions. The electronic power supply cabinet occupied the middle of this wall. The 440-cycle power supply could be placed almost anywhere since cable length was not a concern, so it was situated on the opposite wall. Amplifier and multiplication cabinets were placed on an end wall near the metering desk. Some of these amplifier cabinets held time-delay elements. The AC Calculating Board, which was used for steady-state problems rather than the transient problems primarily done on the Anacom, was situated in an adjacent room, to facilitate interchange of information between them. Figure 8 shows scenes from the construction of the Anacom.

Improvements were made to the Anacom over the years.[49] For example, switches were changed several times. The first switches were noisy, motor-driven mechanical devices, which were replaced in 1953 by switches using vacuum tubes. Around 1970, changes were made to increase capacity and speed up the set-up time by building in prewired transmission line sections with three-phase plugging to replace old spring-retractable cables that were copies of the AC network calculator plugs used in the 1940s and early 1950s. The old cables were used to build "pi sections" to ease the setup of transient problems. At the same time a better arrestor model was built in, using advanced electronics. With these changes, the machine was renamed Anacom II.

The next major change, to Anacom III, occurred in 1980. (See Figure 9.) As the general manager responsible for Anacom at the time recalls the change:

> The change to Anacom III was around 1980. By then statistical/probability measures had been introduced into the art of insulation coordination, and it was necessary to make many runs with random breaker operating times to get the probability distributions of the overvoltages. This was manpower intensive and therefore expensive. [The fastest way to do the calculation was by having] Anacom run by a digital control computer which could work through the night unattended. So we added an HP [Hewlett-Packard mini-] computer to control the switches and to gather the data. Cameras were eliminated, and the computer was programmed to plot the results, even the probability distributions, report-ready. That was Anacom III.[50]

The Anacom was designed to be a general-purpose calculating device, but with a configuration that enhanced its utility in applications of interest to Westinghouse. It was originally conceived mainly to do calculations associated with transient analysis for the electric utilities, electrical transients that arise in the design of generators and motors, applied mechanics problems such as vibration analysis, and problems associated with the statics and dynamics of beams. Over time, the number of applications grew, expanding beyond these original categories to a wide array of applications. The accuracy of one percent to five percent was satisfactory for most applications.

Figure 8. The Anacom under construction.

Figure 9. The Anacom III.

183

In the direct-analog method all elements of the physical system, or nearly all, have their counterparts in the analog and as such are under the direct and independent control of the operator. To study the effect of varying a certain element of the system, the operator need only vary the corresponding element of the analog. In rare cases this may involve a lessened mathematical generality and the use of a few more elements. Some of the advantages of a direct 1-to-1 correspondence between analog elements and system elements in design work are quite apparent. However, not until one enters the nonlinear domain of saturation, of fluid flow, of discontinuous functions, of hysteresis, and of limiting action does the great advantage become truly evident. Rapid and accurate solutions of complex nonlinear problems can be obtained readily.[51]

The Anacom quickly became the most important computational tool of Westinghouse and was used by almost every manufacturing division. During the first sixteen months of operation, 141 problems of 43 distinct types were solved using it.[52]

The Analytical Department did a wide range of calculations soon after Anacom became available in 1948. Using Anacom and other calculating machines, the Analytical Department did calculations for the design of the blades for steam turbines, the study of transient torques in the shafts and couplings of machines, lightning and switching transients in power systems, and the regulating systems (mentioned above in the discussion of Harder's dissertation) for steel mills, paper mills, and voltage regulators. Anacom was used both for internal product development and for contract work for outside organizations.[53] Anacom was also used in consulting projects for regional utility companies, such as the design of Virginia Electric Power Company's first 500-kV system, which was one of the first networks for which it was determined that switching transients were an important system design parameter.

One calculation done in 1948 deserves particular mention: the design of the large compressor blades for the Tullahoma Wind Tunnel in Tennessee. The problem involved the determination of the blade's higher modes of vibration. Because the design problem required knowing third and fourth modes of vibration, and it had only been possible to calculate the first two modes of vibration, designers had previously resorted to building and testing successively larger models—an expensive and labor-intensive process. R. H. MacNeal, working with the Cal Tech analog computer, developed an analog of a cantilevered beam, and Harder's group set this problem up on the Anacom. Because the differential equation representing the cantilevered beam had a closed-form solution, this problem provided a test of Anacom's accuracy. When the test was satisfactory, Harder's group modified MacNeil's analog, tapering the beam and twisting it into the shape of a turbine blade. Although there was no closed-form solution for the equation representing the blade, Harder assumed the validity of the analog and used Anacom to make the Tullahoma design calculations. Once the analog was set up, it was easy to apply an audio oscillator to it and measure modes of vibration and natural frequencies. In 1949, when Westinghouse acquired an IBM 602A for its accounting department, Harder's group redid the calculations numerically and showed Anacom's accuracy on this calculation to have been within two percent.[54]

The importance of calculation to Westinghouse increased after the war, and Anacom was supplemented by digital computers. The company acquired an IBM Card-Programmed Calculator (CPC) in 1952 and an IBM 650 in 1954. In the 1950s the company also added a PACE electronic differential analyzer (a small, relatively inexpensive machine manufactured by Electronics Associates, Inc., which was based on the design of the wartime Bell Labs Mark-9 gunnery computer used to shoot down buzz bombs). Thus by 1955 Harder's Analytical Department contained the DC Calculating Board, the AC Network Analyzer, the Anacom, the PACE electronic differential analyzer, the IBM CPC, and the IBM 650. The company later added additional digital computers: an IBM 704 in 1956, and later an IBM 7090, an IBM 7094, and several Cray supercomputers.[55]

The digital computers first supplemented and later supplanted the analog devices for many problems. The analog machines were faster than the CPC, but the latter received increasing use because its card-programming feature allowed programs to be as long as one liked and to be run by a technician-operator rather than a skilled engineer. The IBM 650, with its magnetic drum memory, was powerful enough that it began to be used for power network problems that for twenty-five years had been handled on the AC Network Calculator. Regulating system problems, which were originally solved on the Anacom, were moved to the PACE machine. After 1970, when the electronic differential analyzer was simulated by the MIDAS (Modified Integration Digital-Analog Simulation) program, most regulation problems were carried out on digital computers simulating electronic differential analyzers. The DC Calculating Board was then no longer used for power system problems, but was applied in the 1950s to diffusion and heat flow problems that could be represented in the form of the Laplace or Poisson equations. The DC Board was particularly useful in the design of early nuclear reactors, to calculate the diffusion of fast flux and slow flux fields.

Anacom survived in the digital computer era.[56] One of its great virtues was that, once an analog was set up, everything could be measured immediately. It was used in the 1950s, for example, for oil flow problems, bearing lubrication problems, diffusion problems, nuclear reactor design, and many other kinds of problems represented by differential equations. As digital computers improved rapidly during the 1950s, progressively more problems were shifted from the Anacom to digital machines. Anacom was sometimes used in tandem with a digital computer; for example, a network solution would be done on the Anacom and the corresponding stability study on the digital computer. After 1970, digital computers were used for all problems except transient, lightning, and oscillatory problems and nonlinear problems of electric power systems. By the 1980s, most lightning problems were also solved digitally. Advertising literature for the Anacom indicated that, in 1990, the machine was still being used for the study of electrical transient phenomena and for the analysis of industrial power systems, such as "power networks that supply arc furnaces where switching transients and resonances are of primary concern."[57]

The Anacom had its greatest value to Westinghouse and its customers, who benefited from the increased efficiency of design and the enhanced productivity it provided. Anacom was also the model for a number of other analog computers. Westinghouse built a copy for J. F. Calvert's laboratory at Northwestern University, which used it mainly for armament studies under contract to US Army Ordnance.[58] Several machines similar to the Cal Tech analog computer were built by McCann's company, Computer Engineering Associates, for the aircraft industry.[59] Similarly designed machines were built by CEZI (the Edison group in Italy) and ERA (a British engineering firm). Altogether, there were probably twenty or thirty Anacom-like machines around the world.

Professional Activities

Harder extended his influence beyond Westinghouse through his involve-ment in the professional community. His knowledge of analog computing led to his appointment to the American Institute of Electrical Engineers (AIEE) Computer Committee, on which he served throughout the 1950s and early 1960s, and whose subcommittee on analog computers he chaired. In the early 1960s he served as the chairman of the AIEE Science and Electronics Division. When the AIEE and the Institute of Radio Engineers (IRE) merged to form the Institute of Electrical and Electronics Engineers (IEEE) in 1963, Harder was primarily responsible for undertaking the merger between the AIEE Computer Committee and the IRE Professional Group on Computers—a tricky assignment because the two organizations had different philosophies for handling their computer special-interest groups. AIEE elected a small group of twenty people to organize its activities, while IRE allowed any of its members who had an interest in computers—thousands of engineers—to join its special interest group. Harder became a Fellow of AIEE in 1948 and served on the AIEE Board of Directors from 1960 to 1962. He was a member of the IEEE Board of Directors from 1966 to 1968.

ASPRAY: *If you were to point to some things that had industry-wide or profession-wide value, what specific things that you contributed would you point to?*

HARDER: *Well, I think the professional society activities would fit best in that category: The contributions to IEEE, IFIP, and AFIPS all were certainly industry-wide things. It's a way that we have of unconsciously sharing large developments. In the IEEE all of the current contributions end up in papers and are presented for everybody to know about. They are the inspiration and starting point for the next group of papers. So in a sense we are carrying out this development of electric power systems, or whatever you're working on, jointly. All the engineers are working jointly and contributing their efforts to this pool. We're pooling it, teaching each other and advancing much faster than if we ever tried to keep that all secret to ourselves. So it's this cooperative development through joint sharing of things in the industry-wide IEEE and similar organizations that is responsible for the huge development of electrical engineering and computers and electronics and all that in this country. If we didn't have that, it would never have developed as fast as it has. So I think that my part in the IEEE, IFIP, and AFIPS contributed to this overall advance more than my specific contributions in just what I did in Westinghouse.*[60]

Harder was also active in the International Federation of Information Processing (IFIP), the international computer organization that arranged for multinational glossaries of computer terms, did basic standards work on ALGOL and other programming languages, formed study groups on administrative data processing and computers in medicine, and much more. Harder himself was responsible for IFIP forming in 1969 its Committee on Computer Applications in Technology, which studied the "research, design, manufacture, operation, and control of products and physical systems, as well as the related programming methods."[61] He believed in the importance of IFIP's work:

One of the greatest ways for advancing our profession, and the industry in which we all contribute and make our livelihood, is the professional technical interchange provided by our technical societies. Here each advance is brought to the attention of our peers, is discussed, criticized, added to, and soon inspires the next advance. Here we meet new friends, though possible competitors in industry, and share in the advance of our profession.

In the various countries of this world this interchange is provided by member societies, the member dues largely financing the operation. However, in the international federations of societies, such as IFIP, ways have also been found to finance an international professional interchange, for the benefit of the whole world. The secret lies partly in the fact that a tremendous international interchange can be effectively organized and carried out at a tiny fraction of the total cost of travel, time, and services provided by all the participants. That small cost can be provided for by very carefully managed budgets, with the limited income from national dues, congress and conference surpluses, and royalties from publication.[62]

Harder orchestrated this financial plan while he served as the chairman of the IFIP Finance Committee from 1966 to 1969 and as IFIP Treasurer from 1969 to 1972. He also served as US Representative to the IFIP General Assembly and made all the US arrangements for the first IFIP Congress, held in Munich in 1962.[63]

Harder's early involvement with IFIP led him to be serve as president from 1964 to 1966 of the professional society that represented the United States in IFIP, the American Federation of Information Processing Societies (AFIPS). AFIPS had only three original members, AIEE, IRE, and the Association of Computing Machinery (ACM). During his presidency, Harder began the sensitive business of expanding the membership, screening out inappropriate organizations, accepting as members the Linguistic Society and the Society for Computer Simulations, and beginning the process of reallocating the revenue of the Joint Computer Conferences to the member societies. He also oversaw the Joint Computer Conferences at a time when the computer industries on both sides of the Atlantic and the Joint Computer Conference exhibits were expanding rapidly.

As a result of his technical and professional achievements, Harder was the recipient of several awards. In 1962 he received the AIEE's Lamme Medal "for meritorious achievement in the development of electrical machinery." In 1971 he received the AFIPS Distinguished Service Award "in recognition of his outstanding contributions to the federation, his guidance and leadership as AFIPS President, and his many contributions to the international growth of the computing profession." He was awarded the Westinghouse Order of Merit, the company's highest recognition. In 1990 he received the first and only Lifetime Innovation Award given by Westinghouse, which cited "his 66 issued patents and his technical achievements as instrumental in increasing the corporation's stature in engineering and innovation" and his work as "a strong force in the development of young engineers as designers of electrical apparatus."[64] He was elected to the National Academy of Engineering in 1976. He has also been a member of CIGRE (*Congress Internationale pour le Grande Reseau Electrique*, an international organization dealing with large electrical systems), the Association for Computing Machinery, the American Mathematical Society, and the National Society for Professional Engineers.

Consultant and Author

In 1965, at age 60, Harder decided to retire from Westinghouse. The Analytical Department was running smoothly and he had been doing the job for twenty years, so he thought it was time for a change. Retiring was not as easy as he had anticipated. Almost immediately upon giving word of his retirement, Monteith called him downtown to corporate headquarters and persuaded him to continue at Westinghouse as a senior consultant.

The company had a broad licensing agreement with Siemens, but Monteith worried that Westinghouse was not gaining all that it could from this agreement. Over the next five years, Harder made fifteen trips to Germany to visit the Siemens laboratories and talk with their engineers. He uncovered in Germany many technical ideas of value to Westinghouse, and he was able to stimulate a number of Westinghouse engineers to visit Siemens and exchange technical information.

I would uncover things that were of interest. I remember in the nuclear field, I found out that Siemens had a way of putting boron into and out of the water by an equilibrium process. At one temperature and pressure it would dissolve in the water, and at another temperature and pressure would go out of the water. I got a consulting engineer from the Nuclear Department to go along with me. He was taking notes galore. It turned out that the boron in the water is a shim. You have rods that control the reactor, but to get the starting point, the boron is a shim to where you want to be operating. Siemens's reactors were all baseload. They had no need to change. But they were developing this in the laboratory. They understood it and were developing it. Apparently our fellows had never heard about it. So I took him over, and we put it right into use in our reactor, and it saved two or three hundred thousand dollars over a job. To get the density of boron down, we didn't have any way of taking it out of the water and putting it back in. So that was probably the most spectacular of the things I learned.[65]

At the end of five years, Westinghouse was forced by legal considerations to sever its long-standing licensing agreement with Siemens. General Electric had a broad licensing agreement with Germany's other large electrical manufacturer, AEG. When Siemens and AEG formed joint venture companies to manufacture large electrical equipment, Westinghouse felt obliged to sever the relationship with Siemens because the US government was already scrutinizing Westinghouse and GE for antitrust activity, and this back-door connection through Germany increased the appearance of antitrust activity between them.

Harder was ready in any event for "retirement" so that he would have time for his own research and writing interests. Many of the 150 papers Harder published during his career presented broad overviews of technical areas.[66] He chose a similar objective for his principal retirement writing project: a book intended to give an overview of the nine sources of energy in the world (coal, oil, gas, nuclear, hydro, wind, solar, geothermal, and biomass).[67] During the several years he devoted to research on this book, he visited energy installations across the United States and western Europe. The book described the characteristics of each of the nine sources of energy, plus the overall resources, physics, chemistry, transportation, and storage of energy. He considered such issues as world population and world food consumption as energy issues.

Conclusions

Many historians believe that the most challenging and interesting intellectual work occurs during the formative years of an engineering discipline. For this reason, historians of power engineering have mainly concentrated on the period up to the 1920s, by which time the fundamental elements of power systems had been determined and power grids had been laid in most urban areas.[68] Harder's career shows, however, that many challenging and interesting problems remained in the 1930s, 1940s, and 1950s—especially in the regulation and control of power systems. Some of these problems involved the need for additional control as power grids were interconnected and increased in scale; others concerned the regulation of new power machinery (generators and motors) and systems incorporating them (steel mills, paper mills, and so on).

Harder's mathematical ability, and his deft application of this talent, set him apart from many of his power engineering colleagues. Mathematical analysis was fundamental to his inventive process as well as to his later work managing the Analytical Department. While a cursory examination of his career might uncover little connection between these two stages of his career, a closer examination shows a strong thread of continuity involved increasingly with mathematical modeling and computational tools.

Harder did not set out consciously on a career in computing. His strong college training in mathematics was simply part of the standard electrical engineering curriculum at Cornell. He chose to work at Westinghouse because it was one of the largest power engineering firms, and in the company's General Engineering Department because of his interest in systems work. Only incidentally was he drawn into mathematical modeling and the use of analog and digital computing devices as a result of some of the systems problems he was assigned. Impressed by the effectiveness of these computational techniques, he continued to use them in his engineering assignments. It is easy to envision how, if contextual factors such as the Depression and internal restructuring within Westinghouse had been slightly different, Harder might have continued as a productive inventor of electrical power components and systems and not been drawn into the computing field.

Harder has acknowledged the Anacom as his most important contribution. Although he did not conceive the machine, he made such fundamental contributions to the design and construction of the full-scale version and to the management of its operation for twenty years, there is no question that he is the person who made computing an important activity at Westinghouse. The methods Harder developed for the Anacom's use, the professional staff he trained, and the cooperation he established with other Westinghouse departments enabled the Analytical Department to be an important internal resource for the company throughout most of the post-war period—continuing long after Harder's own retirement.

Anacom's value to Westinghouse is easy to visualize, though hard to quantify. It was particularly helpful in the company's internal design process and in solving difficult systems problems for customers. It is somewhat more difficult to assess the Anacom's importance outside Westinghouse. Over time, perhaps as many as thirty general-purpose analog computers resembling Anacom were built world-wide. Many of these were direct spin-offs of the Anacom, built either by Westinghouse or by McCann's firm. It is hard to determine, however, what their overall significance was.

It is perhaps useful to compare Anacom with the better known differential analyzers built by Vannevar Bush.[69] What accounts for the difference in recognition accorded these machines?[70] It is not the capabilities of the machines that distinguish them: Each could handle a wide range of abstract differential equations, and each could and did find application to problems associated with voltage regulators, servomechanisms, and nonlinear electrical and mechanical systems.[71] Nor does the number of machines built of each type provide an explanation: There were no more than fifty differential analyzers built around the world, compared to approximately thirty Anacom-like machines.[72] One reason may be that Bush's machines were built earlier than the Anacom, when there were fewer powerful, general-purpose computers, and they were used for important military applications during World War II. Perhaps more important was that Bush's machines had a high profile at MIT, where they were used by many industries, as well as by many students who went on to important computing careers. Meanwhile, Anacom and Anacom-like machines were used predominantly by engineers at Westinghouse and a few other power engineering firms.[73]

The staying power of the Anacom was remarkable, in part because of its capacity to model a complete power system. Bush's differential analyzers, which could at most represent a few circuits at a time, were all taken out of service by the early 1960s. The most nearly contemporaneous digital computer, ENIAC, which was less than two years old when Anacom was placed in operation, was decommissioned in 1955. Anacom was still valuable enough in the 1980s to be purchased by a leading engineering firm, Asea Brown Boveri, and it continued in operation until 1991. Beginning around 1954, one after another application previously carried out on Anacom or other analog computers in the Analytical Department was transferred to digital computers, but even in 1990 there remained classes of nonlinear power engineering problems for which analog machines were orders of magnitude more effective.

There are probably many other reasons why the Anacom lasted so long. Harder mentioned the fact that he built the Anacom with nice cabinets and quality components, so that it had the look of a permanent machine rather than temporary experimental apparatus. The durability and reliability of the equipment meant that it did not become a maintenance problem as it aged. The equipment was soon depreciated, and the company's money was tied up instead in its trained staff, which was already familiar with the analog approach. The organizational structure at Westinghouse, with a single Analytical Department containing both digital and analog computing devices managed by people with analog experience, also meant that managers recognized the value of analog techniques and did not have to make a choice between analog and digital.

Whatever other factors may come into play in explaining the longevity of Anacom, there is no doubt considerable truth in the statement that Anacom continued to be used because it effectively met a need in a historically neglected, but nevertheless important computer application area. Both Anacom and Harder deserve greater recognition than they have so far received.

One of Harder's great loves is sailing. Over the past forty years, he has built and repaired his own boats and made numerous boating trips, including repeated visits to his favored spots on the Georgian Bay in Canada. Thus in September 1991, at age 86, he drove from Pittsburgh to the Georgian Bay to begin another solo sailing voyage. The local newspaper picks up the story at this point:[74]

"His trip progressed reasonably well until he encountered some rough weather at Vidal Island to the east of Meldrum Bay. At this point he decided to anchor on the calm side of Crescent Island and having put out an anchor he laid down to take a rest. The anchor however did not hold and the boat moved into the rocks on Crescent Island, which is on the north side of Vidal Island.

"Mr. Harder then tried to set another anchor to hold the boat but when he was attempting to throw it out a wave hit the boat throwing both the anchor and Mr. Harder into the cold water. The water was not deep and he was able to get back on board.

"There was about 10 inches of water in the boat at this time but there was dry clothing that he was able to change into.

"He stayed on the boat for Tuesday and Wednesday and then thinking Vidal Island was the main Manitoulin he set off in his little dingy. The outboard motor on the dingy had been damaged but he took a set of oars, four slices of bread and a jug of water.

"When he arrived on Vidal Island he realized that it was not the Manitoulin so he made himself a shelter, got a white flag up and waited for help.

"Help did not come and at 3:00 a.m. on Friday [night, actually Saturday morning] he decided that he would try and get to Meldrum Bay on his own. He did have five layers of clothing on for warmth but his feet were extremely cold so he decided it was time to move.

"It was a clear night, the moon was bright and wind had subsided.

"It was not until 10:00 p.m. on Saturday evening that he finally arrived at the dock

in Meldrum Bay in his small dingy. Juhani Paronen was at the dock at the time and he could hardly believe his eyes when this elderly man rowing arrived at the dock in this little boat. After some 18 hours on the water in this small craft Mr. Harder had to be helped out of the boat and into Mr. Paronen's car."

Harder was not seriously harmed by the incident and wrote matter-of-factly to his relatives about the incident. He was able to salvage his boat, the Cozy Cub, which he was planning to repair for a return to the Georgian Bay in September 1992. Perhaps most indicative of Harder, however, were the few lines of calculations he scribbled at the bottom of the article he sent me about this incident:

11 mi. x 5280 = 58080 ft.

25 stroke/min (out of 18 hr. — say 15 hr. of rowing, 900 min.) = 22,500 strokes

58,080/22,500 = 2.58 feet/stroke

---

[1] Gwen Bell, Founding Director of the Computer Museum, Boston.

[2] In 1991, while this paper was in preparation, ABB Power Systems took the Anacom out of operation and released the staff.

[3] This paper would not have been possible without the full cooperation of Edwin L. Harder. The author conducted an extensive oral history interview with Dr. Harder 30–31 July 1991 as part of the IEEE Oral History Program supported by the IEEE Life Member Fund. This interview is the basis of the biographical and much other material in this paper. Dr. Harder generously made available for my use an unpublished article entitled "The Anacom," provided photographs, and answered numerous questions. I also wish to give my thanks to David Morton, who served as the 1992 IEEE Life Member Fund Summer Intern at the Center for the History of Electrical Engineering. Mr. Morton helped with the analysis and writing of the material on Dr. Harder's patents in the electric power field. My colleagues, Andrew Goldstein, Frederik Nebeker, and Eric Schatzberg, provided useful criticism of early drafts of this paper. My graduate assistant, Jill Cooper, helped in many ways with the collection of historical information. An earlier and shorter version of this article appeared in *Annals of the History of Computing*, vol. 15, no. 2, 1993.

[4] Harder is the second of four generations of electrical engineers. His son, William Harder, is an electric power engineer who has worked for Westinghouse and American Electric Power Service. His grandson, Steven Harder, is an electrical engineering graduate from the University of Dayton.

[5] ROTC training involved summer training at a military installation. Most of the ROTC electrical engineering students at Cornell were assigned to the Signal Corps at Fort Monmouth, New Jersey. However, Harder was assigned to the Ordnance Department at Aberdeen Proving Ground, presumably because of his professed interest in mechanical engineering.

[6] Interview 1991, pp. 128–129.

[7] Harder indicated that he preferred systems work to the work of the other departments, which mainly built apparatus. He seemed to enjoy the overview afforded by systems work and the technical problem-solving challenges this department confronted.

[8] The total business of Westinghouse was $216 million but then dropped

precipitously due to the worsening economic climate. The Pennsylvania Railroad contract was the one bright spot. This order affected not only the East Pittsburgh plant, which built the locomotives but also the Sharon, Pennsylvania plant, which built the transformers and the Newark plant, which built the relays.

[9] The calculations themselves were somewhat routine inasmuch as the company had already conducted similar calculations for the Virginian, Norfolk, and Western and the New York, New Haven, and Hartford railways. The Switchgear Department had a DC calculating board for calculations to determine how big the circuit breakers should be and for setting the relays. While this DC board was adequate for 60-cycle systems, it did not give sufficient accuracy for the 25-cycle railroad electrification systems.

[10] There had been litigation involving Westinghouse, the electric utility industry, and Bell Telephone over the telephone interference caused when rectifiers were introduced. Rather than continue to litigate, the parties formed a joint commission to study and correct the situation. Harder's calculations were part of the work of this joint commission.

[11] In his oral history, Harder recounted another instance where his computing skills and knowledge of harmonics came in handy. Long Island Railroad cars traveled into New York City on DC power, while Pennsylvania Railroad cars arrived on 25-cycle AC and the signal power system employed 100-cycle AC. With the AC and the DC trains on the same track, some of the DC got into the AC transformers, which produced unusual harmonics of the 100-cycle AC and sometimes caused the signals to fail. In one instance, two trains approached head-on in a tunnel in New York City but fortunately were able to stop before crashing. Harder helped carry out the calculations that caused the railroads to change the signal power to $91\frac{2}{3}$-cycle AC to avoid any further intermodulation.

[12] Interview 1991, pp. 47–50.

[13] In connection with this work, Harder wrote two chapters in *Electrical Transmission and Distribution Reference Book*, by Central Station Engineers of the Westinghouse Electric Corporation (East Pittsburgh PA: Westinghouse, not dated). This book was used throughout the world.

[14] The board was built by Bill Parker in the Switch Gear Department.

[15] Of course, numerical methods were again used when these problems were moved to digital computers.

[16] See E. L. Harder, "Pennsylvania Railroad New York-Washington-Harrisburg Electrification—Relay protection of power supply system," *AIEE Transactions*, vol. 58, 1939, pp. 266–277.

[17] Harder designed the annunciator systems, the load and frequency control, and the terminal boards for the Hoover Dam switchboards.

[18] See E. L. Harder, "Solution of the general voltage regulator problem by electrical analogy," *AIEE Transactions*, vol. 66, 1947, pp. 815–825, for a publication giving these results.

[19] The magnetic amplifiers were fast enough to replace standard Rototrol amplifiers in the initial stages of the systems, thereby reducing them to the two-delay systems Harder had shown to be inherently stable. During World War II, Harder designed one of the earliest magnetic amplifiers (a balanced

bias, saturated core amplifier). Gordon Brown of the MIT electrical engineering faculty had his students build this amplifier. When the war ended, Westinghouse found that it did not need this new amplifier for its commercial operations. The name "magnetic amplifiers" came into use only after the war. Harder was the chairman of the AIEE committee that developed standard nomenclature and other standards for magnetic amplifiers. Harder's design so impressed Gordon Brown that he invited Harder to join the MIT electrical engineering faculty, an offer that Harder declined for personal reasons.

[20] An oscillograph was used on the Cathedral of Learning at the University of Pittsburgh to measure the instantaneous lightning stroke current as a function of time, but it was too expensive to use at all fifty sites. A less expensive device, called the klydonograph, was used for these. To measure crest lightning current, researchers had used magnetic links, 1½ in. strips of iron placed in the field of the stroke current in a lightning rod or a tower leg. The residual magnetism was a measure of the current. In the klydonograph, about 200 of these links were placed on the periphery of a wheel, driven by a small motor. The different links coming by the recording head during a stroke gave a rough time recording of the current.

[21] In the early years, when the department had only analog computers, all of the professional employees were electrical engineers because it was important that the people doing calculations understand the applications and how they were physically modeled. When digital computers came to be used in the department, he hired some mathematicians and programmers.

[22] Interview 1991, p. 106, slightly edited.

[23] Interview 1991, p. 27.

[24] I am grateful to Harder for his generous assistance in making accurate the technical description of the HCB relay. My first attempt at describing it was replete with errors. Harder rewrote this material, and I have slightly revised his account.

[25] Interview 1991, pp. 33–35, slightly edited.

[26] The patent application was received 29 November 1937 and filed 1 December 1937. Patent number 2,144,494 was awarded 17 January 1939. The invention is described in E. L. Harder, B. E. Lenehan, and S. L. Goldsborough, "A new high speed distance type carrier pilot relay system," *AIEE Transactions*, vol. 57, 1938, p. 5.

[27] Three-phase directional relays were used because the load current in a good phase might be flowing in the opposite direction from the fault current in a faulted phase. In the three-phase relay, the fault current predominated.

[28] Harder, "An Invention Story," unpublished document, revised September 1992.

[29] E. L. Harder, B. E. Lenehan, and S. L. Goldsborough, "A new high-speed distance type carrier pilot relay system," *Transactions of the AIEE*, vol. 57, 1938, p. 5.

[30] Harder, "An Invention Story," unpublished document, revised September 1992, slightly edited.

[31] The patent application was received 4 November 1940 and filed 3 May 1941; patent number 2,331,186 was awarded 5 October 1943. The invention is described in E. L. Harder, et al., "Linear couplers for bus protection," *AIEE*

Transactions, vol. 61, 1942, pp. 214–249.

[32] The linear coupler then became a mutual reactance, and the voltages were totalized.

[33] The patent application was received 1 December 1953 and filed 29 December 1955; patent number 3,027,084 was awarded 27 March 1962. E. L. Harder, "Economic load dispatching," *Westinghouse Engineer* (November 1954): pp. 194–200; "Electrical network analyzers," *Journées Internationales du Calcul Analogique*, September 1955; "Automatic dispatch pays off," *Electrical World*, 18 April 1955; Harder et al., "Loss evaluation, Parts 1–2," *AIEE Transactions*, vol. 73, 1954.

[34] The computer could take into account a whole set of factors that affected economic dispatch: cost curves, cost multipliers, upper and lower limits on production at stations, flow in tie lines, and more.

[35] Interview 1991, p. 158.

[36] For a description see Karl L. Wildes and Nilo A. Lindgren, *A Century of Electrical Engineering and Computer Science at MIT, 1882–1982* (Cambridge MA: MIT Press, 1985). These devices are described briefly in two standard survey histories of computing: Allan G. Bromley, "Analog computing devices," chapter 5 in William Aspray, ed., *Computing Before Computers* (Ames, Iowa: Iowa State University Press, 1990); Michael R. Williams, "The analog animals," chapter 5 in *A History of Computing Technology* (Englewood Cliffs NJ: Prentice-Hall, 1985).

[37] See W. R. Woodward, "Calculating short-circuit currents in networks— Testing with miniature networks," *The Electric Journal*, August 1919, pp. 344–345. A general description of analytical solutions to electric power network problems is given in Robert D. Evans, "Analytical solutions," *The Electric Journal*, August 1919, pp. 345–349.

[38] If the impedances are represented as $R + jX$, in these railroad calculations the ratio of X/R was small, which meant that the DC Board could not provide accurate calculations.

[39] See Harder, chapter 10, "Steady state performance of systems, including methods of network solution" in *Transmission and Distribution Reference Book* (East Pittsburgh PA: Westinghouse Electric Corporation, 1942).

[40] See H. A. Travers and W. W. Parker, "An Alternating-Current Calculating Board," *Electric Journal*, May 1930, pp. 266–270.

[41] Copies of the AC Calculating Board were built by Westinghouse for external use. For example, Westinghouse built one for Gibbs and Hill, the engineering firm doing the engineering of the Pennsylvania Railroad system. Many were sold to utilities all over the world.

[42] Travers and Parker, p. 268.

[43] See R. D. Evans and A. C. Monteith, "System recovery voltage determination by analytical and A.C. calculating board methods," *AIEE Transactions*, vol. 56, 1937, p. 695.

[44] See H. E. Criner, G. D. McCann, and C. E. Warren, "A new device for the solution of transient vibration problems by the method of electrical-mechanical analogy," *Journal of Applied Mechanics*, September 1945, pp. A135–A141; G. D. McCann and H. E. Criner, "Mechanical problems solved electrically," *Westinghouse Engineer*, March 1946, pp. 49–56.

45 See E. L. Harder and G. D. McCann, "A large-scale general-purpose electric analog computer," *AIEE Transactions*, vol. 67, 1948, pp. 664–673, for a list of references to these applications of electric analog machines. See also the description of Anacom in G. D. McCann and E. L. Harder, "Computer—Mathematical Merlin," *Westinghouse Engineer*, November 1948, pp. 178–183.

46 Harder and McCann, p. 664. The Anacom was intended to have the capacity of the electrical and mechanical transient analyzers, the servo-analyzer, and extra facilities—additional multipliers, transformers of special design, forcing functions of the independent variable, and nonlinear functions of the dependent variable.

47 The two computers were "alike in all essential features and general design, differing only in minor matters of construction and in the exact number of each kind of element installed." (Harder and McCann, p. 664) The differences between the two machines are listed in an appendix to the Harder-McCann article. The other machine was known as the California Institute of Technology Electric Analog Computer.

48 There were some differences in the configuration of the Westinghouse and Cal Tech models of the machine based upon their intended uses. See the appendix to Harder-McCann for a discussion of these differences. The Cal Tech machine was not designed, as was the Westinghouse machine, for use in conjunction with an AC Calculating Board. Some of the use of the Cal Tech machine was directed at aeronautical rather than electric power applications.

49 The information about changes to Anacom was supplied in conversation with Harder and in a letter from C. J. Baldwin, Manager, Engineering Projects, Asea Brown Boveri Power T&D Company to author, 29 June 1992.

50 Baldwin to author, 29 June 1992.

51 E. L. Harder and J. T. Carleton, "New Techniques on the Anacom—Electric analogy computer," *AIEE Transactions*, vol. 69, 1950, pp. 547–556 (citation from p. 547).

52 Ibid.

53 Harder was clearly proud of his operation. He claimed that the top engineers in the company gravitated to his operation and that in his twenty years managing the operation his engineers did not make an error using the Anacom that caused a serious problem for any Westinghouse department. Part of the reason was his insistence that his engineers all have the big picture and not blindly accept results that came from the machine. Another was his insistence on good computer usage practice. "One of the bad uses [observed in one of the Westinghouse manufacturing departments] was cases where programmers tried to program something for a division and the thing just didn't work that way. They didn't understand how a shop works. They imposed the structure. They assumed that each workman when he finished his day would be willing to sit down and write down what he did and put it on punch cards and you could keep track of everything. Those fellows weren't hired to do that. They would no more do that than fly. And so the complete program was a flop" [Interview 1991, p. 104].

54 For the 602A, the blade problem was set up as an iterative calculation. The blade was divided into ten portions and the vibrations were calculated

through the blade at different frequencies. It took two weeks of calculation—putting cards through the 602A and sorting them, over and over again—to obtain a curve of frequency versus response. Harder reports that "one good creative engineer resigned in this process. He felt that there must be a better way to make a living than standing in front of a machine for 8 hours a day, for two weeks, feeding and sorting cards, in order to get one answer. (He later became editor of the *Westinghouse Engineer*.)" ["The Anacom," unpublished manuscript.]

[55] Harder did not consciously choose to remain with IBM equipment once there were good alternatives on the market. The Westinghouse accounting department made the decision and paid part of the cost of the 7090. After that, no one in the company was willing to pay the expense of rewriting all the administrative software to consider an alternative vendor. Harder was, however, pleased with the computers and service from IBM.

[56] The strengths and weaknesses of digital and analog calculators for power applications, as of 1945, is considered in E. L. Harder, "Electrical network analyzers," International Analog Computing Meeting, Brussels, Belgium, September 1945, *Proceedings*, pp. 419–433.

[57] ABB Power Systems, Inc., Advanced Systems Technology, "Anacom III," descriptive bulletin TR-IJ-COM. The brochure lists typical transient problems done with the Anacom III: insulation requirements, surge arresters, switching surge reduction, torsional interaction, performance of static VAR compensators, series capacitor applications, operating procedures, circuit breakers, transients caused by faults, equipment design, ferroresonance, shunt reactors, and arc furnace supply circuits.

[58] Calvert was a good friend of Harder; they had shared a desk in their early days at Westinghouse. Calvert had turned down the opportunity to found a magazine with a friend at the end of World War I to pursue an engineering career. The magazine, which succeeded beyond anyone's imagination, was *Reader's Digest*. Calvert had a distinguished career, eventually heading the electrical engineering department at the University of Pittsburgh [Interview 1991, p. 148].

[59] McCann hired H. E. Criner, his collaborator at Westinghouse on the mechanical transients analyzer computer, to head Computer Engineering Associates. They eventually had a falling out and the company was sold to Susquehanna Corporation. Almost all of its staff, the company's only asset, left Susquehanna's employ within a year.

[60] Interview 1991, pp. 156–157, slightly edited.

[61] I. L. Auerbach, "IFIP—The early years: 1960-1971," in Heinz Zemanek, ed., *A Quarter Century of IFIP* (Amsterdam: North Holland, 1985), p. 87.

[62] E. L. Harder, "Financing IFIP: 'The things we can do together that we can't do separately,'" pp. 335–336 in Zemanek (cited above).

[63] On the former point, see AFIPS Board of Directors Minutes, 17 November 1969, Charles Babbage Institute Archives.

[64] Paul Lego, quoted in Interview 1991, p. 156.

[65] Interview 1991, pp. 83–84, slightly edited.

[66] See, for example, his chapters on "Steady-state performance of systems including methods of network solutions" and (with J. C. Cunningham) "Relay and circuit breaker application" in *Electrical Transmission and Distribution Reference Book* by Central Station Engineers of the Westing-

house Electric and Manufacturing Company, (Chicago: Lakeside Press, 1942); also (with W. E. Marter of Duquesne Light) "Principles and practices of protective relaying in the United States," *Transactions of the AIEE*, vol. 67, 1948, p. 1005.

[67] *Fundamentals of Energy Production* (New York: John Wiley, 1982).

[68] For example, Thomas Hughes, *Networks of Power* (Baltimore: Johns Hopkins University Press, 1983), the leading history of power engineering, ends in 1930. Richard Hirsh, *Technology and Transformation in the American Electric Utility Industry* (Cambridge, UK: Cambridge University Press, 1987) explores the later history of power engineering but characterizes much of it as a period of stasis.

[69] It is much more difficult to compare Anacom with a digital computer of its era, such as ENIAC. The digital computers led directly to commercial products, which were on the market in large numbers by the end of the 1950s. Neither differential analyzers nor Anacom-like machines were extensively manufactured. Most of the analog computers sold commercially in the postwar period were small electrical analog machines based upon the gunnery computers built during World War II. These devices were not nearly as powerful as the Anacom.

[70] Perhaps the difference in recognition was not as great at the time as the historical literature now suggests. Harder was recognized by his computing peers, as his election to professional positions in IFIP, AFIPS, and AIEE indicate.

[71] McCann-Harder, p. 182. In an unpublished article on the Anacom, Harder did identify some differences in operating principles between the two machines:

"*Electric Analog Computers* are of two general types, passive and active. The Anacom is basically a passive element computer, but did have amplifiers to represent the gains in regulating systems (no longer used).

"The active system, using operational amplifiers as integrators, adders, etc., is an electronic differential analyzer, the counterpart of the mechanical differential analyzer of Vannevar Bush."

[72] In any event, the number of technological artifacts does not necessarily correspond to the technology's importance.

[73] It is interesting to note, however, that the MIT machines were built mainly for electric power applications, as was the Anacom. The historical literature on computing gives little attention to power engineering as a driving application for the design of computing equipment—perhaps because this literature emphasizes digital computers.

[74] "86 year old man ship-wrecked 5 days," *Manitoulin Recorder*, Wednesday, 25 September 1991.

CHAPTER
7

New Applications
of the Computer
Thelma Estrin
and Biomedical Engineering

THELMA ESTRIN

Figure 1. Thelma Estrin, former president of the IEEE Engineering in Medicine and Biology Society, former executive vice president of the IEEE, recipient of the Distinguished Service Citation and an honorary doctorate from the University of Wisconsin, winner of the IEEE Haraden Pratt Award, and current resident of Santa Monica, California.

At the beginning of 1942 the United States faced a severe labor shortage. Even before the US's entry into the war a month earlier, factories were expanding production and working overtime to make this country, in President Roosevelt's phrase, "the arsenal of democracy." In the first months after the attack on Pearl Harbor, millions of men left their jobs to enlist in the military, and the labor shortage became acute in the defense industries as the production of aircraft, ships, trucks, artillery, uniforms, munitions, and other matériel increased dramatically. The government concluded that the only solution was to "employ women on a scale hitherto unknown."[1] An advertising campaign, featuring Rosie the Riveter, urged women to serve their country by entering the work force. The campaign was successful—by the middle of 1945 the female work force had increased by 6.5 million, a 57 percent increase.

One woman who helped meet this need was an 18-year-old New Yorker who enrolled in a 3-month engineering-assistant course at the Stevens War

Industries Training School—recently established by the Stevens Institute of Technology in Hoboken, New Jersey—and then worked for two years at Radio Receptor Company. (See Figure 2.) The young woman went on to a distinguished career as an electrical engineer. Though wartime circumstances influenced the training and employment she took, her determination to pursue a career of her own predated the war.

Figure 2.   Thelma Estrin in the machine shop of the Radio Receptor Company of New York City.

From Schoolgirl to Engineering Assistant

Thelma Austern was born 21 February 1924 to Jewish parents in New York City's Harlem section, then mainly a middle-class area.[2] Her father, I. Billy Austern, was a shoe wholesaler and traveled a great deal. Her mother,

Mary Ginsburg Austern, was unusually independent for a woman at that time. Before her marriage she ran an automobile parts store; she also drove a car, which few women then did. After her marriage she gave most of her energy to maintaining a kosher home and rearing her only child, but continued to be active in the local Democratic Party and in the Order of the Eastern Star (a fraternal and service society associated with freemasonry).

The Depression brought hard times to the family. They moved to Brighton Beach in Brooklyn, first to a fairly nice apartment, then to a less expensive one. It was always Mary Austern's dream to own a piano and have her daughter take lessons, but this was never possible. Mary's hope that her daughter excel in her education was, however, fulfilled. She expected Thelma to go to college and follow a profession, preferably law, and she did not want her to study typing as many other girls then did.

Thelma did well in all subjects and liked mathematics especially. At Abraham Lincoln High School she took an extra math class during her senior year, partly through the influence of her friend Richard Bellman, who later became a famous applied mathematician (and recipient of the IEEE Medal of Honor). Thelma and Richard dated for some time, but went their separate ways. One factor was a difference in political orientation: Thelma was at the time active in the American Student Union, a liberal group, while Bellman was more conservative.

It was through the American Student Union that Thelma met Gerald Estrin in June 1941. He was a history major at the City College of New York; that January, Thelma had begun studies there. On 21 December, two weeks after the attack on Pearl Harbor, she and Gerald were married. It was also in that period that both of her parents died: Mary Austern of cancer in March 1941 and Billy Austern of polycythemia in January 1942.

The war caused Thelma and Jerry, like countless others, to change their plans. Jerry enlisted in the Army Signal Corps and, before being called to active duty in 1943, worked at Kurman Electric Company, which made electromagnetic relays. Thelma, as we have seen, enrolled in the engineering-assistant course at Stevens Institute of Technology and then worked at Radio Receptor Company in New York City. A company publication reported, "We have seen Thelma manipulating lathes, shapers and surface grinders and squinting at micrometers and surface gauges with all the aplomb of an experienced toolmaker."[3] The publication reported also that history and tennis were among her favorite hobbies (commenting, "From a casual survey of her physique, we would venture to guess that she spends more time with a racquet than with Beard and Prescott") and that radio was her chief interest. In accord with this interest, she was soon transferred from the tool and model shop to the company laboratory, where she assembled test equipment and repaired radio transmitters. She also took evening classes in engineering at City College.

In 1945 Thelma joined Jerry in Montgomery, Alabama for a short period before they were moved to San Bernardino, California. There Thelma worked as a radio technician for the Army Air Force. When the European phase of the war ended, Jerry was transferred to Salt Lake City. Thelma returned to New York City and resumed studies in engineering at City College. Mechanical drawing was the major challenge. She says she had little aptitude for three-dimensional visualization, and "Where most students would work for a half hour, I would work three times as long."[4] But she finally gained facility at mechanical drawing and, in later life, taught the subject.

When my husband Jerry was a senior in high school (1936), he was very interested in becoming an engineer, particularly electrical, because he was outstanding in mathematics (and disliked chemistry). However, it was commonly recognized that Jewish engineers would rarely be hired by large industrial corporations, particularly power engineering firms. In fact, many private universities would not allow Jewish young men to major in engineering, because they supported industry's biased point of view. City College in New York City did allow all qualified young men to enroll in engineering, but even in the radio and electronics field employment possibilities were very limited. Jerry therefore decided to go to the City College School of Business and Management, where we met. With the second world war, the electronics industry bloomed and Jerry entered the Signal Corps.

When Jerry obtained his Ph.D. in electrical engineering, in 1950, he was strongly recommended to Bell Telephone Laboratories by his professor, but Bell did not readily hire Jewish engineers. Jerry was interviewed and might have received a position if he hadn't decided to accept an offer from John von Neumann at the Institute for Advanced Study. Interestingly, RCA employed many electronic engineers who were Jewish (David Sarnoff who headed RCA was a Jew), but to our knowledge Bell employed no Jewish engineers.[5]

An EE Education at an Accelerated Pace

Jerry was discharged from the Army in December 1945, and three months later he and Thelma moved to Madison, Wisconsin, to enter the undergraduate electrical engineering program at the university. They managed financially because Jerry had GI Bill support, Thelma sold her mother's diamond ring, and both got part-time work as teaching assistants. Thelma reports, "At school nobody took me very seriously. Most of my classmates thought that Jerry was either keeping me in school to keep me out of mischief, or so that I could help him with his homework. But I took myself seriously and Jerry took me seriously. From the beginning, our marriage . . . has been a partnership. This was something that came to both of us quite naturally. I was an only child whose mother wanted me to be a professional woman. . . . Jerry's mother was a small business woman, so neither of us had to overcome the mindset of believing women's place was only in the home."[6]

They both had EE experience as well as college credits, and they were both extremely industrious. "By working 18 hours a day and not taking vacations, we both zipped through Wisconsin, obtaining B.S., M.S., and Ph.D. degrees in record time."[7] Thelma received her B.S. in 1948, her M.S. in 1949, and her Ph.D. in 1951. As a graduate student, Jerry received research assistantships—this helped him complete his Ph.D. a year earlier than Thelma—while she could get only teaching assistantships. She believes that this was the result of the prevalent attitude that her interest in engineering would last only until she and her husband had children.

Estrin, however, was determined to maintain a career, and she wanted to do this while still having children and raising a family. She therefore decided to concentrate on analytical rather than experimental EE, as she believed that analytical engineering would better permit interruptions and part-time work than would experimental engineering.[8] She asked Professor Thomas J. Higgins, an authority on numerical methods as employed in electrical engineering, to be her advisor. Her master's thesis was an extension of a known method (double Laplace transformations) to solve problems in electric circuit analysis and electromagnetic theory, and she coauthored with Higgins an article presenting her main results.[9] In her doctoral thesis she showed how to use the method of incremental areas to calculate, to any degree of accuracy, the charge distribution of a capacitor consisting of two planar plates.[10] Here too she coauthored with Higgins an article presenting her main results.[11]

As they were nearing completion of their Ph.D.s, Thelma and Jerry received identical telegrams from Bell Aircraft offering a job at $5000 a year. Jerry, however, had a much more exciting opportunity: to join John von Neumann's computer project at the Institute for Advanced Study in Princeton, New Jersey.

To Princeton and into Medical Electronics

Three papers written jointly by von Neumann, Arthur Burks, and Herman Goldstine in the years 1946 to 1948 presented for the first time a detailed description of an electronic stored-program computer, and in 1946 von Neumann began a project to build such a machine at the Institute for Advanced Study (IAS).[12] Von Neumann originally estimated that three years would be required to complete a functional machine. It turned out that it took six years, and in the summer of 1950 much work remained.

As soon as Jerry completed his Ph.D. in June 1950, the Estrins moved to Princeton. For three or four months Thelma also worked on the IAS computer project—to test and document the arithmetic unit of the machine. She then returned to her dissertation research. This involved complex and lengthy calculations, for which she drew on local resources— the simultaneous equation solver (an early analog computer) of the Radio Corporation of America, and the wife of an Institute physicist armed with a Marchant calculator.

When Estrin received her Ph.D. in the summer of 1951, she sought employment. Because she didn't want to work at the same place as her husband, almost the only possibility in the Princeton area was the research laboratory of the Radio Corporation of America. Though she did receive an interview (with Jan Rajchman), she was not hired by RCA. She later said, "RCA was hesitant, and I'm sure it was because I was a woman."[13]

Estrin finally found a position in New York City, at the Electroencephalography Department of the Neurological Institute of Columbia Presbyterian Hospital, and began work there in November 1951. The EEG department consisted of a clinical laboratory and a research laboratory. Estrin later wrote: "I was responsible for the reliability of the clinical EEG equipment and supervised a technician who maintained the equipment. I was also given the opportunity to do research and to collaborate with physicians on EEG and EMG [electromyography] studies. At that time I was as much a rarity among engineers for my technical work as I was for my sex, for biomedical engineering had not yet become a field of engineering."[14] One of her first tasks was to take over the development of a frequency analyzer for bioelectric potentials; she improved the circuit design to increase the stability and ease of tuning of the device.[15] She also collaborated on a study of the action potential and refractory period of striated muscle.[16]

Despite the job in New York and the two-hour commute each way, Thelma was part of the social life of the Institute for Advanced Study. The Estrins were good friends with Jule and Elinor Charney—Jule was directing the effort to use the IAS computer for numerical weather forecasting—and with Julian Bigelow, chief engineer for the computer project. Bigelow, aware of Thelma's work at the EEG department, encouraged her to consider using an electronic computer to do statistical analysis of brain waves (something she did in later years). Thelma and Jerry went out to dinner many times with Johnny and Klari von Neumann. Johnny always questioned Thelma in detail about her work on the electrical activity of the nervous system, because of his interest in the relation between the computer and the brain.[17]

Israeli Interlude

In December 1952 Gerald Estrin was asked to be part of an effort to build an electronic computer at the Weizmann Institute of Science in Rehovot, near Tel Aviv.[18] At that time plans were underway to build versions of the IAS computer at a dozen locations in the United States and Europe.[19] The Israeli effort received full support—including a complete set of plans—from the leaders of the Institute project, von Neumann, Bigelow, and Herman Goldstine.

The departure for Israel was delayed by the arrival of the Estrins' first child, Margo, in February 1953. Thelma had stopped work at Columbia shortly before. In September Jerry took a leave of absence from the Institute project, and he and Thelma, with a baby in tow, spent three months in Europe, visiting computer groups in England, Netherlands, France, and Italy. In late December they reached Israel.

When he arrived at the Weizmann Institute, Gerald Estrin was surprised to find that he was to be director of the computer project. Apprehensive because of the lack of equipment and materials in Israel at that time and because of the need to recruit and train staff, he nevertheless accepted the challenge. An enthusiastic team was gradually assembled, and by dint of hard work and resourcefulness the supply and fabrication difficulties were overcome one by one.

Except for one month—at the time of the birth of a second daughter, Judith, in November 1954—Thelma was a principal member of the engineering group. (See Figure 3.) Though the computer, known as WEIZAC (for WEIZmann Automatic Computer), was closely modeled on the IAS computer, significant redesigns were necessary, and Thelma played a large role in this process.[20]

Figure 3.   Mechanical assembly of the WEIZAC chassis. Thelma is at the right in the white lab coat.

After fifteen months in Israel the Estrins returned to Princeton in the spring of 1955. At that time the WEIZAC central processing unit and primitive input-output were complete; it performed its first calculation six months later and thus became the first electronic computer in the Near East. The time spent in Israel made Thelma and Jerry, neither of whom was religious, identify more strongly with Judaism; their daughters all speak Hebrew and maintain certain Jewish traditions. They returned to Israel for a year in 1963 and have visited the country almost every year since.

When we came to Israel to build this computer in 1954, everybody said to us, "Why does a little poor country like Israel have to spend its money on a computer?" When we left and they gave my husband a certificate which says, "For building the first electronic computer in Israel," I can remember thinking, the first! How many electronic computers do they think there are going to be here, or in Israel. We still, at that time, in 1954, did not realize the potential of it, aside from it being . . . a giant calculating device. A scientific computer for big [numeri-cal calculations], that we knew from modeling problems. . . .We didn't see its explosion into the business world, resource planning, [and] government use to keep track of data. I, at least, did not see that.[21]

In April 1955 the Estrins arrived back in Princeton, and Jerry resumed work on the Institute computer project. Thelma, though offered an engineering position at the Moore School of Electrical Engineering at the University of Pennsylvania, chose to work closer to Princeton as a mathematics instructor at Rutgers University in New Brunswick. In 1956 Jerry was hired by UCLA (as an associate professor in the School of Engineering and Applied Science and the Institute for Numerical Analysis) to initiate a program in computer engineering, and the family moved to Los Angeles. Nepotism rules prevented her from being employed by the School of Engineering, so she found a half-time teaching position at Valley College, a junior college in Los Angeles. (See Figure 4.) She taught there for two years and also did some work as a consultant. A third daughter, Deborah, was born in December 1959.

Data Processing Laboratory at the UCLA Brain Research Institute

In the mid-1950s H. W. Magoun, professor in the UCLA Medical School, initiated an interdisciplinary neuroscience program at UCLA, and this led in 1959 to the establishment of the Brain Research Institute, dedicated to study of the brain using methods from anatomy, physiology, chemistry, and biophysics. The following summer Thelma, because of her work at Columbia on electro-encephalography, was hired by the Brain Research Institute (BRI) to organize a conference on computers in brain research, which was held that fall.[22]

At that time very few biomedical researchers thought of the electronic computer as a research tool. One who did was Mary A. B. Brazier, a physiologist specializing in electroencephalography who was just then joining BRI. Another was a professor of anatomy at BRI, W. Ross Adey, M.D. and ham radio operator. It was at Adey's urging, and with the collaboration of the director of the BRI, Dr. John D. French, that a proposal to establish a data processing laboratory at BRI was drafted and sent to the National Institutes of Health (NIH). At that time local computing capability was a new concept; at universities almost all computing was carried out at a single facility. The grant application was entitled "The application of computing techniques to brain function," and the engineering aspects of that application were the work of Estrin.[23]

Thus in 1961 NIH (specifically, the National Institute of Neurological Diseases and Blindness) provided funding to set up the Data Processing Laboratory (DPL) of the Brain Research Institute. This was the first grant ever awarded by NIH for setting up a computer facility in a medical school, and DPL was probably the first computer facility established expressly for nervous system research. DPL was both a research laboratory for the development of computing techniques for neuroscience and a data processing facility for members of BRI. Because about half of the 90 or so research projects under way at BRI involved recording of electrical signals, it was clear to Estrin that there were many potential users of electronic data processing.[24] The NIH funding for DPL continued until 1981, the annual allotment usually exceeding a quarter million dollars.

Figure 4. Estrin teaching engineering drawing at Valley College in Los Angeles.

The initial NIH grant included funding for Estrin's proposal: to design and implement an analog-to-digital conversion (ADC) system.[25]   The typical electrical behavior of a neuron, as shown in a microelectrode recording, is a series of "spikes" (rapid rises and falls of electrical potential), and neurophysiologists hypothesized that the nervous system encoded information in the temporal pattern of the spikes. (See Figure 5.) Testing this hypothesis required statistical analysis of large amounts of data, a task ideally suited for a computer, but only if the analog recordings were first converted into numerical digits. Estrin, taking advantage of her experience with the digital logic of the WEIZAC, devised a system to do this automatically.[26] The system, incorporating a clock pulse and a count register, digitized the time interval between spikes and produced a magnetic-tape output that could then be analyzed by a computer. Because physiologists at that time seldom had immediate access to a computer, Estrin showed how to obtain, with a relatively simple, self-contained electronic system, real-time analysis of the firing patterns, which could be useful in monitoring an experiment.

Figure 5.   On the left is a microelectrode recording of a neuron, showing a series of spikes. On the right is an electroencephalograph, obtained by placing electrodes on the scalp.

With an electroencephalogram (EEG), which records electrical potentials on the scalp (see Figure 5), neurophysiologists wanted to digitize both amplitudes and time intervals. Estrin designed and built an appropriate ADC system that could be used either to convert an analog tape to a digital tape or to convert analog signals from recording instruments to digital signals for immediate input to a computer.[27] Analog-to-digital conversion is required in almost every use of computers by biomedical researchers (since data are recorded by analog devices and since digital computers require numbers as input), and Estrin's was one of the first ADC systems for biomedical data.[28] Ross Adey used this system for his pioneering work in spectral analysis of electroencephalograms.[29]

In 1962 Estrin obtained a Fulbright Fellowship to spend most of 1963 at the Weizmann Institute in Israel investigating the EEG patterns associated with epilepsy.[30] Because of difficulties with laboratory facilities, Estrin was unable to make much progress on this, but she did continue work on the design of systems for analog-to-digital conversion of physiological data.[31] She returned to the United States and the Brain Research Institute at the end of the year.

By the mid-1960s almost all universities in the United States had a computing center, and almost all of these centers consisted mainly of a large computer operating in "batch mode," which is to say that the computer executed one program at a time and users waited several hours or a day to retrieve their output. In this regime, debugging a program was extremely time-consuming, and some applications, requiring immediate response, were not possible. One solution to these difficulties was to have a computer dedicated to a single application. Indeed, a principal mission of BRI's Data Processing Laboratory was providing such "on-line computing." As so-called minicomputers became widely available in the early 1970s, more and more researchers adopted this approach.[32] Estrin took the initiative at BRI, acquiring for the Data Processing Lab a DEC PDP-12 minicomputer in 1970 and advising three neurophysiology labs to acquire their own PDP-12s.

Another solution, also introduced in the 1960s, to the problems posed by the batch processing regime was "time sharing": many users at remote locations have simultaneous access to the same computer, and each has the illusion of access to a dedicated computer. In 1965 Estrin designed a time-sharing system that permitted data processing in real time at a number of BRI laboratories.[33] As mentioned above, she had earlier designed a self-contained device to give simple data-analysis in real time, but now, for the first time, the experimenter could obtain sophisticated analyses (carried out by computer) as the experiment was being conducted. Not long after Estrin returned to UCLA following the Fulbright year in Israel, she went to work in the laboratory of Mary Brazier, and it was in Brazier's lab that Estrin built an on-line analog-to-digital system that time-shared the SDS 930 computer in the Data Processing Laboratory.

The impact of computers on EEG studies became apparent in the late 1960s. The study of "average evoked potentials" provides one example: Sensory stimuli cause changes in the EEG, but these changes are, in general, detectable only with the signal-averaging techniques made possible by on-line computers. Another example is the proliferation of studies of the alpha rhythm. This is a fairly regular wave, eight to thirteen cycles per second, which is the most pronounced of the rhythms exhibited by EEGs. The alpha rhythm itself was discovered decades earlier, and it was also known that waves of alpha frequency could be induced by repeated flashes, but the computer greatly facilitated the study of both of these phenomena. In the late 1960s Estrin collaborated with John S. Barlow on a study of both intrinsic and induced alpha rhythm.[34] They compared the phasing of the induced alpha waves with that of the intrinsic waves to cast light on the question of possible generating mechanisms.

It was also in this period that Estrin made use of the computer to present EEG information in a new form. A traditional electroencephalogram measures the electrical potential at a small number of electrodes placed on the scalp and displays potential differences for particular pairs of electrodes. Estrin recognized that recent technological advances in analog-to-digital conversion, in computing, and in graphic display made possible spatial display of EEG patterns. This could be done by increasing the number of electrodes on the scalp, digitizing the data, using a computer to complete the pattern by interpolation, and using a computer to generate a graphic display. Moreover, this scheme allowed one to generate a succession of EEG maps, to photograph the maps with a movie camera, and then to project an "EEG movie" that showed the temporal sequence of spatial patterns. (See Figure 6.)

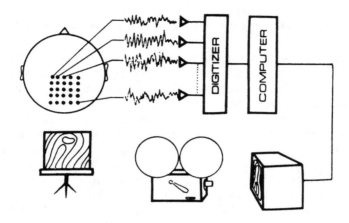

Figure 6. A diagram illustrating the method of generating "EEG movies."[35] The signals from the array of electrodes are amplified and digitized; the computer, through interpolation and map-projection techniques, forms a contour map of electrical potentials, which is displayed on a cathode-ray tube; a movie camera, synchronized with the CRT displays, photographs successive displays; and the resulting film is projected on a screen.

Collaborating with programmer Robert Uzgalis, Estrin solved the many problems required to make this scheme work.[36] Many considerations were involved in finding an appropriate map projection (so that the curved surface on which the electrodes are placed is represented with little distortion on a planar surface) and in devising interpolation routines. The rate at which the camera took pictures had to be controlled by an interface between computer and camera, so that one photograph is taken for each pattern displayed on the CRT; here Estrin and Uzgalis had the assistance of John Whitney, a pioneer in computer animation. Whitney was at that time using DPL equipment to produce one of the first films of computer animation.

In the period from 1965 to 1970 Estrin worked in the laboratory of Mary

Brazier—purchasing, designing, calibrating, and maintaining a variety of laboratory equipment—while also playing a large role in activities of the Data Processing Laboratory. (See Figure 7.) In the late 1960s DPL, which was without a director, encountered increasing budgetary and operating difficulties, and there was doubt about whether grant requests for continued funding would be approved. In 1969 the director of the Brain Research Institute, John French, asked Estrin to become acting director of DPL and to head a task force charged with making recommendations about how the lab ought to be organized.[37] French later wrote, "Dr. Estrin accomplished the task asked of her with industry, determination, skill and tenacity. I am confident the task force would have been unsuccessful save for these qualities of leadership which she exhibited."[38] The reorganization was carried out, and a $300,000 funding proposal, with Estrin as principal investigator, submitted and approved.

Thus in 1970 Estrin became DPL Director, supervising six to ten computer professionals and deciding how to allocate personnel and material resources to the many BRI projects. During the 1970s DPL, which earlier had concentrated on medical research, directed an increasing part of its effort to health care delivery.[39] In this period Estrin offered each year the course "Electronics for Neuroscience"; in addition, she and a colleague gave a graduate seminar on "Computer Applications in Health Care Delivery." She also edited a special issue of the journal *Computer* on "Information Systems for Patient Care."[40]

In the mid-1970s Estrin pioneered the use of interactive graphics as a tool for neuroscientists and neurosurgeons. She collaborated in this work with half a dozen other researchers, especially with Robert Sclabassi and Richard Buchness. In a 1974 paper, Estrin, Sclabassi, and Buchness described a computer system that combined diagnostic information from x-ray scanning of the patient's head with general neuroanatomical information from brain atlases to compute and graphically present a brain map scaled for that particular patient.[41] This work, which drew on the very recent development of computerized axial tomography (itself made possible by high-speed computers) and which made use of a large-screen IMLAC graphics processor interfaced with a DEC PDP-12 computer, could not only generate an individualized brain map, but also simulate and display the movement of an instrument as it is taking place during an operation. Estrin did further work to make this system useful in planning an operation (to allow simulation of the operation taking different trajectories to the targeted structure) and to adapt it to animal stereotaxic surgery.[42]

Estrin, Sclabassi, and Buchness collaborated on another application of interactive graphics, the analysis of the series of spikes produced by a neuron.[43] It is typical for neurophysiological experiments to generate enormous data sets. The ability to present data in graphic form can be

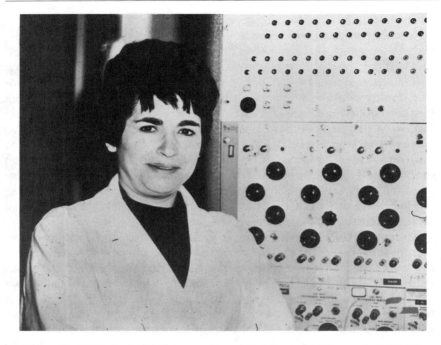

Figure 7. Estrin in front of some equipment that was part of an analog-to-digital conversion unit at the Data Processing Laboratory of UCLA's Brain Research Institute.

extremely helpful, both in providing insights into the data itself and in facilitating the comparison of experimental data and model predictions. Estrin and collaborators demonstrated this for spike-train data by designing and implementing a system that consisted of an IMLAC graphics terminal, a PDP-12 computer, and an IBM 360 computer.

Other work by Estrin in the 1970s included the use of computers to apply sophisticated statistical techniques to the analysis of EEG recordings (in sleep research) and the use of computers in a retrospective study of the efficacy of particular surgical treatments.[44] She also arranged for a long-distance connection of computers: the IMLAC terminal at DPL connected by telephone lines to the Burroughs B6700 at the Davis campus of the University of California, 500 miles away. (This was in order to make use of a computer language, MUMPS, developed specifically for medical applications.)[45]

In the mid-1970s personal computers, then usually called microcomputers, began to be widely used by researchers. Estrin worked to speed this process, especially by showing how microcomputers could be used for data acquisition, which in neurophysiology typically involved considerable preprocessing (such as filtering, artifact removal, digitization, and discrete-event sensing).[46] As Director of DPL, she provided computer support for a variety of research projects and helped dozens of researchers make use of computers (including studies of sudden infant death syndrome, multiple sclerosis, epilepsy, and evoked responses).[47]

In 1979 the NIH grant that had supported DPL for eighteen years was approved but not funded (the overall appropriation for NIH grants that year was lower than expected). There was also less support within BRI than Estrin had hoped. She notes that neuroscience as a whole has moved toward chemical and molecular-biological studies, and that investigation by means of electrical measurements has not been as vigorously pursued, partly because few researchers have had the requisite training or inclination, partly because some of these approaches have proved especially difficult.[48] Also, the Brain Research Institute required additional space, so DPL ceased to exist. Estrin remains optimistic about the prospects of computers in biomedical research, pointing to the growing body of successful computer-aided research and to such recent technological advances as the personal workstation, standard software, fiber-optical devices for data transfer over computer networks, and ever-more-powerful supercomputers.[49]

The Data Processing Laboratory had, however, already had a large impact on neurophysiological research. In the first years of the Brain Research Institute, investigators were generating large amounts of data, typically in visual form, that were assessed mainly by the naked eye; in 1980 BRI investigators were generating more data, most of it in digital form, that were processed and analyzed largely by computer programs. The Data Processing Laboratory of the Brain Research Institute had, along with the Communications Biophysics Group of the MIT Research Laboratory of Electronics, pioneered the use of computers for neurophysiological research. More than a hundred neurophysiological researchers had consulted with DPL staff, and many more were alerted to possible uses of computers by DPL activities.[50] Particular DPL contributions were made in the areas of analog-to-digital conversion, data acquisition and preprocessing, signal analysis, interactive graphics, modeling and simulation, distributed computing, and laboratory computer systems.

Medical Informatics

Though most of her work involved the use of computers in biomedical research, Estrin was also a proponent of computer use in medical treatment and hospital administration. She argued, ". . . clinical medicine is inherently an information processing activity and the clinician is a decision maker who depends on the quality of the data stored in the medical record. . . . Very little funding has been allocated for research on the storage and retrieval of patient information."[51] She believed that computerizing medical records could lead to improved patient care, more efficient administration and accounting, and the facilitation of epidemiological studies and evaluations of medical treatment.[52] Even as computers improved the quality of health care, she argued, they could lower costs and thus extend the reach of such care.[53]

Thus Estrin was an early champion of medical informatics—the application of computers to medical research and treatment—in all its branches. Estrin, however, did not have much success in generating interest at UCLA as, for example, in her effort beginning in 1964 to interest leaders of the medical school in exploring the application of computers to patient records.[54] Nor did she have success in finding a position for herself at UCLA that would allow her to develop medical informatics in a concerted way. In 1976 she was asked to be "coordinator of activities"—a temporary, part-time position—for a nascent institute for biomedical engineering, the Crump Institute. Estrin wanted a greater commitment on the part of the university to her role in forming this institute, and when this was not forthcoming she declined the offer.[55]

In 1980 Estrin became professor in residence in the Computer Science Department of the School of Engineering and Applied Science. (Nepotism rules at the university had been abolished.) That year she introduced a new course on computers in medicine, and the following year a new course on computer literacy for nonscience majors. (See Figure 8.) In her own work she began to give more attention to the use of computers for medical histories, patient management, and clinical decision making.

Figure 8. Estrin teaching at UCLA.

In the early 1980s Estrin served as advisor to Donna Hudson, a UCLA graduate student in computer science. A large part of Hudson's dissertation research was the design of an expert system for medical decision making. Called EMERGE, the rule-based expert system was designed to be machine-independent and to be capable of running on a microcomputer. Because the cardiology division of the UCLA medical school had recently completed a study of decision making in the treatment of chest pain, Hudson and Estrin chose this area for the initial application of EMERGE, producing an expert system for use by emergency room personnel.[56] As part of this work, Hudson and Estrin devised a procedure for deriving rule-based knowledge from medical audits.[57]

Administrative Work at the National Science Foundation and at UCLA Extension

In 1982 Estrin accepted a two-year rotating position at the National Science Foundation, that of division director for Electrical, Computer, and Systems Engineering, and she moved to Washington 1 October 1982. There she was responsible for oversight of more than 400 grants totaling some $30 million annually, and she established a new program, Bioengineering and Research for the Handicapped. Bioengineering had been part of a program that included automation and sensing systems, while research for the handicapped had been a separate, small program; Estrin believed both were strengthened by placing them together in a new program.

A big issue that came up [when I was at NSF] was, should there be a computer science directorate? Nobody who was at NSF wanted another directorate, because it would impinge on their directorate funds. Differences in our division occurred on whether grants were computer science or communication engineering. The program manager for Computer Engineering only had two million dollars in funds, and needed an increase. The manager for Electrical and Optical Communications had about six million dollars in funding, and thought all of the network research were communication issues. There was a Division of Computer Research in the Directorate for Mathematical and Physical Sciences headed by a young man whom I liked very much, but he has died. He did not want to change the structure, nor did the manager of the Information Science and Technology Division in the Directorate for Biological, Behavioral, and Social Sciences. The three directors with computer research funds gave [NSF Director Edward] Knapp a talk on the subject. I did give reasons to support the other directors, but I was leaving in a few weeks and didn't really give my point of view, to bring all of computer research together.

As soon as Erich Bloch arrived at NSF, he formed the Computer Science Director-ate. ECSE became Electrical and Communications Systems, which is the name today. I was pleased by the change. While I was at NSF I was treated very well, and I loved working there and being in Washington. We sublet an apartment in Watergate, with a view of the Potomac. It was great. Jerry joined me for the second academic year.

. . . When you have one person who has been the director of a program for many years, that program director will have funded a large group of the same people over those years. They visit these people, are abreast of their work, use them as reviewers for new grant applications, et cetera. They often become friends with the people they support. These funded people become well known in the field and are liked for what they have accomplished. Then a new young investigator applies for funds, and money is very tight. They just don't get the same kind of reviews, unless their original research is absolutely super.

NSF does sponsor very superior research, probably fifteen percent of the time. The rest of the funding is high quality . . . . In parts of engineering and in computer science the field changes every two to three years. An NSF program director (or manager) may know the field very well, but may not be up to date with new emerging areas. They may not get new reviewers because they have their own reviewers, frequently professors whom they also fund, and whom they know will return a review quickly. There are contradictions in the present system of reviews, but in general, I think it's a pretty fair system . . . . It is important to change directors of programs and divisions. That is why the rotating director from academia is a good idea.[58]

While at NSF, Estrin lobbied strongly for funding for the first grant in 1984 to the National Technological University (NTU)—a university in which all courses are given by video by means of C- and Ku-band satellite transmission. NTU made it possible for people, many of whom had no access to other forms of advanced technical education, to earn master's degrees. This experience and her conviction that "lifelong learning . . . has to become a way of life for the engineer" (because technology is changing so fast)[59] made her interested in a position in continuing education at UCLA— Director of the UCLA Extension Department of Engineering and Science. She held this position, serving also as assistant dean of continuing education, from 1984 until 1989.

In some respects Estrin found the work satisfying. The UCLA extension program is one of the largest in the country—the department Estrin managed offered some 550 courses to more than 15,000 students each year—and is highly regarded.[60] She worked to increase the number of courses conducted at high-tech companies, such as Northrop and Rockwell. She introduced courses in artificial intellegence, manufacturing engineering, and optical communications and a program in telecommunications and communications engineering. And she worked to revamp a master's program between Extension and the School of Engineering.

In other respects, however, Estrin found the work frustrating. The main problem was that bureaucratic inertia made it almost impossible to be entrepreneurial. She wanted to introduce videotaping of courses and to establish a library of videotaped lectures. She wanted to provide remote instruction with lectures conveyed by satellite.[61] She wanted to set up an institute in the South Bay to offer courses mainly for aerospace engineers. She wanted to turn some certificate programs into external master's programs. But University Extension was a large, bureaucratic organization, and proposed changes faced the additional hurdle of approval by the University as a whole. Moreover, because the extension program had to be self-supporting, it was almost impossible to find money for new ventures. So Estrin was unsuccessful, despite her best efforts, in bringing about these new activities.

In her last years of teaching, 1989/90 and 1990/91, Estrin introduced two new courses, one on technology and society (concerned both with the nature of engineering and the role of technology in modern society) and one on women in engineering, and revamped an undergraduate course on the engineer in society. In July 1991 she retired from UCLA and was named professor emerita.

IEEE and Other Organizations

In the late 1970s Estrin came to feel that the advances in neuroscience made by molecular biologists and biochemists were more significant than the studies of the gross electrical activities of the brain. This was one of the reasons she left the Brain Research Institute and accepted an appointment in the School of Engineering (and there were no longer nepotism rules that forbade this). Another reason was that she began devoting more and more of her time to professional activities, particularly for the Institute of Electrical and Electronics Engineers but also for the Association for the Advancement of Medical Instrumentation and other biomedical societies. These activities involved a great deal of travel and countless hours of arranging things from her office by telephone, correspondence, and, later, electronic mail. In 1981 she said, "[professional activities] is what takes up my time and so I decided brain research and I had sort of come to our end."[62]

Estrin had joined the American Institute of Electrical Engineers (one of the two predecessor societies of IEEE) as a student at the University of Wisconsin in 1948, but she had hardly been active in the Institute until 1973 when she was encouraged to run for the Advisory Committee of the Engineering in Medicine and Biology Society (one of the technical societies within IEEE). She was elected and found this work rewarding, and in 1977 she was elected to be president of the Engineering in Medicine and Biology Society.

In 1979 Estrin ran successfully for the position of director of Division Six of the Technical Activities Board, which encompassed six IEEE technical societies. Division Directors are also members of the IEEE Board of Directors, and she thus became the first woman to be elected to the

Board.[63] In 1982 she was elected executive vice president of IEEE and thus again served on the Board of Directors. She instigated the process to establish the Judith A. Resnik Award, which was finally established in 1986 to recognize an electrical engineer for contribution to space engineering.[64] Aside from remaining active in the Engineering in Medicine and Biology Society and besides her work within IEEE for women in engineering (discussed below), Estrin has served, from the mid-1970s to the present, on about a dozen IEEE committees.[65]

IEEE honored her work in biomedical engineering by naming her a Fellow in 1977 "For contributions to the design and application of computer systems for neurophysiological and brain research" (the sixth woman to attain this honor)[66] and by awarding her a Centennial Medal in 1984. IEEE honored her "for outstanding service to the Institute" by awarding her the Haraden Pratt Award in 1991.

Outside of IEEE Estrin has been active in the Association for the Advancement of Medical Instrumentation (AAMI) and in the Biomedical Engineering Society (BME). She served several years on the Board of Directors for each of these societies, and she was a member of the editorial board for BME's *Annals of Biomedical Engineering*. In the late 1970s Estrin was vice president of the Alliance for Engineering in Medicine and Biology (a group of 23 engineering, scientific, and medical societies in the US founded in 1969 to facilitate cooperation in sponsoring an annual conference), and she was the US member on the administrative committee of the International Federation of Biomedical Engineering (to which the Alliance belonged). In the 1980s Estrin served on two boards of the National Research Council (Telecommunications and Computer Applications, and Energy Engineering) and on several accreditation committees. She has also been active in the American Association for the Advancement of Science (AAAS); she was elected chair of the AAAS Engineering Section for 1989 and chair of the Computer Section for 1994.

In the 1970s Estrin argued for the recognition of a new branch of engineering, clinical engineering, concerned with the development and implementation of technology for health care.[67] In 1974 the AAMI and seven other medical associations established a certification procedure for clinical engineers, with a provision that those who at that time already had considerable experience as clinical engineers could be "grandfathered" into certification (without going through the new procedure in its entirety). The fact that the clinical engineers were virtually all men and the sexist language of the certificate (including "Know All Men by these presents that . . . has given satisfactory evidence that he has met the qualifications. . . .") prompted Estrin to apply for certification. She explained, "My motivation to apply for certification stemmed from my desire to have a woman "grandmothered," as well as my interest in promoting the clinical engineering profession."[68]

In 1978 Estrin was appointed to the Board of Trustees of Aerospace Corporation, becoming the first woman member of that board. Her background in computing was probably a factor in this appointment, as Aerospace was then making special efforts to take advantage of computer technology. She served on the Board's Technical Committee and also the Compensation and Personnel Committees. She also supported the efforts of the Aerospace Women's Committee, which had been formed in 1972, to encourage the hiring and promotion of women. When she retired from the Board in 1982 on assuming the NSF position (a government employee cannot serve on the Board), she was honored "for her wisdom, friendship, and enduring contributions to the Corporation."[69]

Estrin was a member of the Army Science Board from 1980 to 1983 and a member of the NIH Biotechnology Resources Review Committee from 1982 to 1986. The former committee advised the Army on technological matters; the latter provided guidance for the Biotechnology Resources Program, which was established by NIH in 1962 "to provide support for complex technological capabilities in biomedical research."[70]

Women in Engineering

At many points in her career, Estrin encountered discrimination, though she was determined not to allow the prejudices of others to diminish her drive to achieve certain goals. There were a few occasions when she felt discriminated against because she was Jewish and spoke with a strong New York accent. She faced a second type of discrimination at the Brain Research Institute: discrimination against "technicians." Many BRI members saw a sharp distinction between scientists and technicians (including all engineers): The former do the important intellectual work, while the latter merely take care of technical problems. When told that Estrin had won a major award, a scientist who knew her registered surprise and commented, "But she's just a technician!"

Much more serious in Estrin's estimation, however, is discrimination against women. "I don't think we fully appreciate how difficult it is to overcome a socialization process in which women are brought up to be housewives, and men to be wage earners. As a result, many women lack confidence, have low esteem and fear success. The sexist attitude of men further drains our confidence and our energy and frequently we blame ourselves when we don't achieve our goals."[71] In most professions a woman must overcome both overt discrimination—something Estrin has encountered many times[72]—and the "covert barriers of early conditioning in a dominant-male culture [which] remain as fossilized attitudes."[73]

Though a long-time member of the Society of Women Engineers—she reported on her work at the first SWE convention in 1953—it was in the 1970s that Estrin became active in the effort to increase the number of women in engineering. Her most important work was as chair of the Committee on Professional Opportunities for Women (COMPOW) from 1975 to 1980. COMPOW is a committee of the IEEE that was organized by Julia Apter in 1971. Its purpose, as expressed by Estrin, is to promote the five Rs: recruitment of women as students of engineering, retention of women students enrolled in engineering programs, retraining or continuing education of women who have career interruptions, redress of grievances for women who suffer from discrimination, and re-education of the profession to the suitability and desirability of women as engineers.[74]

I was always in the Engineering in Medicine and Biology (EMB) Group, later a society, but I didn't think much about it. There was a woman by the name of Julia Apter, who was a physician and who died at a young age. She was on the Administrative Committee of EMB in about 1972. She began to write to the few women in EMB, and I responded. She was a very attractive woman who was very hostile with men, particularly those in NIH and in EMB, and was always in a battle with them. She could also be very charming. She was the woman who proposed the Committee for Professional Opportunities for Women [COMPOW] in IEEE.

. . . She wanted me to run for Ad Com [in 1972], and was going to get ten signatures to put me on the ballot. I replied that I was certainly pleased to run, but I'd get my own signatures, and put myself on the ballot, which I did and was elected to EMB Ad Com. I met Julia at my first meeting. She said that the men were never going to allow me to contribute, or would they socialize or invite me to dinner with them. None of this was true. I thought that the Ad Com of Engineering in Medicine and Biology was sincere, friendly, and interested in working with me.[75]

NEBEKER: *There's also the problem that you alluded to, that—at least in some environments—promotion depends on your visibility and your self-promotion, and many men are better at that than women.*

ESTRIN: *You do get some women who are considered very aggressive. They just have to be aggressive to overcome the barriers. Some women overdo that approach and become openly hostile, as Julia Apter, the founder of COMPOW, was. Men are more narrowly focused on the project they're doing than on the interaction with the environment around the project. Men are also more analytical, and women have a very good intuitive approach. Of course, I'm discussing men as a group, not as individuals.[76]*

Estrin provided COMPOW with five years of energetic and innovative leadership. She arranged for hospitality suites at professional conferences in order to facilitate women's networking. According to Estrin, "Professional women who are sent to conventions like NCC [National Computer Conference] by companies have a sense of isolation. They walk into a solidly male environment. Hospitality suites operated by the companies are generally smoke-filled rooms, jammed with a bunch of men. When a woman walks in, they rarely look at her as a professional or a colleague."[77] She established a newsletter for women engineering students, encouraging women to become more active in IEEE. She organized workshops at conferences, both IEEE conferences, such as Electro and Wescon, and other conferences, such as NCC (where in 1978 a workshop was entitled "Designing and Debugging Careers for Women in the Computer Industry").

It was not only through IEEE that Estrin pursued the cause of women in engineering. She has given many talks and published many articles on the topic. At UCLA she gave a course entitled "Women in Engineering," and she was cofounder of the UCLA Chapter of Association for Women in Science. In 1978 she tried unsuccessfully to establish a new organization, Advancement of Women in Technology (AWIT), with the goal of encouraging women to enter technical occupations and promoting upward mobility for women in such jobs. (Very recently an organization with similar goals, Women in Technology International, has been formed.)

Estrin has been a member of Systers, a computer network for female computer scientists. Two of the greatest barriers to success by women in science and engineering are the difficulty women have in finding mentors ("... most women have a tough time getting this important guidance") and the "old-boy network" ("the basic obstacle ... which is 'still very much in place' ").[78] In 1987 Anita Borg and several other computer scientists established a computer network to ameliorate these difficulties.[79] The 900 current members of Systers use it in various ways, such as soliciting career advice, asking who is researching a particular topic, or requesting guidance on writing papers.

Estrin argues that in the modern world, engineering is increasingly done, not by individual effort, but by teamwork. This requires excellent communication skills and the ability to accommodate different ideas, approaches, and workstyles. While men often excel in competition, women tend to prefer a cooperative environment and to work well in larger groups. "Women make excellent managers because they tend to be more interpersonal, more sensitive to people's feelings, and better organized."[80] She argues also that a more realistic picture of electrical engineering—not just a lot of math, physics, and apparatus, but solving practical problems in complex contexts—would attract more women into engineering.

Estrin's efforts were recognized by the Society for Women in Engineering by its Achievement Award, and when the Association of Women in Computing was formed in 1982, Estrin was named an honorary member. In her acceptance speech for the SWE Achievement Award, Estrin briefly reviewed the careers of six outstanding women engineers and remarked, "I would like to tell you that at least one of those women was my role model. But the truth is, I was the first woman engineer I ever knew."[81]

I recall Emily Sirjane, who was an invaluable chief administrator of IEEE, who knew everything about the organization from A to Z. She retired because she was not well and also because of her age, and she did die shortly after that. She was not particularly interested in women, she was interested in everybody in IEEE, and was very pleasant and a wonderful worker. I remember when the Board was writing a farewell to her. They talked about her secretarial skills—she was very neat, and very precise, and she followed procedures—but it was the kind of talk that you would only write of a woman. . . . I recall saying this and mentioning the outstanding administrative leadership she provided. Well, the Board agreed with me and ultimately gave her a much nicer final letter.[82]

From a letter to Emily Sirjane from Thelma Estrin, 25 January 1979:

We both joined the IEEE in 1948; you as a staff member in New York and I as a student member in Wisconsin. . . .

Until the early 70's I harbored the notion, formed in my student days, that the Institute was run by a "group of stuffy old men." That myth was dispelled when I attended my first USAC meeting to consider accepting the leadership of COMPOW. I realized that I had become of an age with the "men" and that they were interesting people. At the time I did not know that behind the IEEE scenes was a wise, talented and beautiful woman, but I quickly found out as I became an active volunteer. Your intimate knowledge of IEEE, your effective implementation of policies and procedures, your wise judgement in making decisions, and your gentle firmness with people have been a joy to witness and work with. I will miss you very much.

Bruno Bettleheim . . . said "We deeply need women scientists and engineers who are committed, as human beings and as good workers, to their profession, and who are committed to it in line with their female genius." You have brought the female genius of a woman manager to IEEE.[83]

Family and Honors

Estrin gives a great deal of credit for her achievements to her husband: "So while I did not have a role model, and didn't have a mentor, I did have a supportive husband."[84] And she is obviously proud of her children, whom she calls "my three greatest contributions . . . joint authored with my husband. . . . "[85] The eldest, Margo, is a physician in private practice, specializing in internal medicine. "She calls herself the oddball of the family because she's the only one who is not an engineer."[86] Judith, a computer engineer with a master's degree from Stanford, is a founder and senior vice president of Network Computing Devices, a company with some 300 employees. She claims that she's the family oddball, because she doesn't have a doctoral degree. The youngest, Deborah, who, uniquely for this family, has not claimed to be family oddball, earned a computer science Ph.D. at MIT. Her speciality is computer networking, and she is now an associate professor of computer science at the University of Southern California. (See Figure 9.) The children became achievers, according to Judith, not because of any pressure to do so, but through the example of their parents' commitment to the work ethic.[87] All three daughters are married and have children.

Figure 9.   A photo of the Estrin family taken in 1981.

Estrin is justifiably proud of the part she has played in the development of the discipline of biomedical engineering and of her work in showing the usefulness of computers in biomedical research and healthcare delivery. For these achievements she has won recognition: To the many honors already mentioned may be added Fellowship in the AAAS for "interdisciplinary research contributions to neuroscience and computer science" and an Honorary Doctor of Science Degree by the University of Wisconsin.[88] She is proud of her daughters and of the fact that her grandchildren will, in her words, "grow up knowing that men and women can be equal partners at work and at home; a very rare concept when I was married. . . . "[89] Perhaps she should be proudest of the fact that her example and her efforts on behalf of women in engineering will extend this concept to a great many besides her children and grandchildren.

[1] Quoted in William Chafe, *The American Woman: Her Changing Social, Economic, and Political Roles, 1920–1970* (New York: Oxford University Press, 1972), p. 137.

[2] The information about Thelma Estrin contained in this article comes mainly from the following sources: (1) an extensive oral history interview of Estrin conducted by the author 24 and 25 August 1992 (from which an edited transcript has been prepared); (2) a large collection of personal papers (referred to as the Estrin Papers); (3) e-mail correspondence between Estrin and the author; (4) Estrin's published writings; and (5) other published writings. (The transcript of the extensive interview, copies of personal documents and e-mail correspondence, and a full list of Estrin's publications are available at the IEEE Center for the History of Electrical Engineering.)

[3] Unsigned "Thumbnail biography" from a 1943 Radio Receptor Company publication (*Radio Receptor News*), Estrin Papers.

[4] Interview 1992, p. 11.

[5] Letter Estrin to Nebeker, 10 December 1992, Estrin Papers. In *Genius: The Life and Science of Richard Feynman* (New York: Pantheon Books, 1992), James Gleick writes (pp. 84–85), "On the eve of the Second World War institutional anti-Semitism remained a barrier in American science. . . . when he [Feynman] was at MIT, the Bell Telephone Laboratories turned him down for summer jobs year after year, despite recommendations by William Shockley, Bell's future Nobel laureate. Bell was an institution that hired virtually no Jewish scientists before the war."

[6] Estrin, "Women engineers—female magicians," *U.S. Woman Engineer*, October 1981, pp. 2–5.

[7] Ibid.

[8] "Plan for study" from a fellowship application submitted by Estrin (probably in 1949) to the American Association of University Women, Estrin Papers.

[9] Estrin, "Theory and application of multiple Laplace transforms to the solution of problems in electric circuit analysis and electromagnetic theory" (unpublished M.S. thesis, University of Wisconsin, 1949), and T. A. Estrin and T. J. Higgins, "The solution of boundary value problems by multiple Laplace transformations," *Journal of the Franklin Institute*, vol. 252, 1951, pp. 153–167.

[10] Estrin, *Determination of the Capacitance of Annular-Plate Capacitors by the Method of Subareas* (unpublished Ph.D. dissertation, University of Wisconsin, 1951).

[11] T. A. Estrin and T. J. Higgins, "Determination of the capacitance of annular-plate capacitors by the method of subareas," *Proceedings of the National Electronics Conference*, vol 20, 1964, pp. 939–944.

[12] William Aspray, *John von Neumann and the Origins of Modern Computing* (Cambridge MA: MIT Press, 1990).

[13] Quoted in Janet Noonan, "Thelma Estrin: computers on the brain," *Westside Women Today* (special supplement to the Santa Monica CA *Evening Outlook*), 26 July 1979, pp. 40–41.

[14] Estrin, "Computers, neuroscience and women: (1949–1999)," *Proceedings,*

Sixth Annual Conference of the IEEE Engineering in Medicine and Biology Society (1984), pp. 831–836.

[15] T. Estrin and P. F. A. Hoefer, "Revised frequency analyzer for bioelectric potentials in the sub-audio range," *Review of Scientific Instruments*, vol. 25, 1954, pp. 840–841. The earlier system is described in C. Markey, R. L. Shoenfeld, and P. F. A. Hoefer, "Frequency analyzer for bioelectric potentials in the sub-audio range," *Review of Scientific Instruments*, vol. 20, 1949, pp. 612–616. A full description of the revised analyzer, including twenty schematics and half a dozen photographs, is in the Estrin Papers.

[16] P. F. A. Hoefer, G. H. Glasre, C. Hermann, Jr., and T. A. Estrin, "Action potential and refractory period of striated muscle in man," *Federation Proceedings*, vol. 12, 1953, item 220.

[17] John von Neumann's Silliman lectures were published in book form as *The Computer and the Brain* (New Haven CT: Yale University Press, 1958).

[18] Gerald Estrin, "The WEIZAC years (1954–1963)," *Annals of the History of Computing*, vol. 13, 1991, pp. 317–339.

[19] Aspray, *John von Neumann*, especially pp. 91–94.

[20] A detailed account of the construction and use of the computer is contained in Gerald Estrin, "The WEIZAC years. . . ."

[21] Interview of Estrin conducted on 9 April and 9 May 1981 by Louise Marshall for Neuroscience History Research Project of the UCLA Brain Research Institute.

[22] The conference proceedings were published in 1961: *Computer Techniques in EEG Analysis*, edited by Mary A. B. Brazier (Elsevier, Amsterdam).

[23] Estrin, "The UCLA Brain Research Institute Data Processing Laboratory," *A History of Medical Informatics* (ACM Press and Addison-Wesley Publishing Company, 1990), pp. 157–173.

[24] T. Estrin, W. R. Adey, M. A. B. Brazier, and R. T. Kado, "Facilities in a brain research institute for acquisition, processing and digital computation of neuro-physiological data," *Proceedings of the Conference on Data Acquisition and Processing in Biology and Medicine* (Oxford: Pergamon Press, 1963), pp. 191–207.

[25] See memo, 10 November 1961 from Brain Research Institute Computer Committee (written by Estrin) to F. H. Sherwood in the Estrin Papers.

[26] Estrin, "Recording the impulse firing pattern of neurons utilizing digital techniques," *Digest of the 1961 International Conference on Medical Electronics,* 1961, p. 99.

[27] Estrin, "A conversion system for neuroelectric data," *Electroencephalography and Clinical Neurophysiology*, vol. 14, 1962, pp. 414–416.

[28] Slightly earlier ADC systems were built by W. M. Siebert and collaborators at the Massachusetts Institute of Technology (see Communications Biophysics Group of the Research Laboratory of Electronics and W. M. Siebert, "Processing neuroelectric data," MIT Technical Report 351, 1959) and by W. R. Uttal and P. A. Roland (see W. R. Uttal and P. A. Roland, "A terminal device for entry of neuroelectric data into an electronic data processing machine," *Electroencephalography and Clinical Neurophysiology*, vol. 13, 1961, pp. 637–640). Slightly later an ADC system for physiologic data was developed at the laboratory of Earl H. Wood at the Mayo Foundation in Rochester, Minnesota (see E. H. Wood, "Evolution of instrumentation and

techniques for the study of cardiovascular dynamics from the Thirties to 1980, Alza Lecture, April 10, 1978," *Annals of Biomedical Engineering*, vol. 6, 1978, pp. 250–309).

[29] W. R. Adey, "On-line analysis and pattern recognition techniques for the electroencephalogram," in G. F. Inbar, ed., *Signal Analysis and Pattern Recognition in Biomedical Engineering* (New York: John Wiley and Sons, 1975).

[30] Application for Fulbright Fellowship, 1962, Estrin Papers.

[31] Estrin, "Analog to digital conversion of physiological data," *Medical Electronics* (Proceedings of the Fifth International Conference, Liége) (Desoer, Belgium; 1963), pp. 638-648.

[32] Estrin, "Minicomputers in biology and medicine," *Proceedings of the Second Jerusalem Conference on Information Technology* (Jerusalem, July 1974), pp. 667–684.

[33] Estrin, "On-line electroencephalographic digital computing system," *Electroencephalography and Clinical Neurophysiology*, vol. 19, 1965, pp. 524–526. See also Estrin, "Neurophysiological research using a remote time shared computer," *1966 Rochester Conference on Data Acquisition and Processing in Biology and Medicine* (London: Pergamon Press, 1968), pp. 117–135.

[34] J. S. Barlow and T. Estrin, "Comparative phase characteristics of induced and intrinsic alpha activity," *Electroencephalography and Clinical Neurophysiology*, vol. 30, 1971, pp. 1–9.

[35] Figure 2 from T. A. Estrin and R. C. Uzgalis, "Computerized display of spatiotemporal EEG patterns," *IEEE Transactions on Biomedical Engineering*, vol. BME-16, 1969, pp. 192–196.

[36] Ibid.

[37] Memo 21 January 1970 from French to Carmine Clemente, and Memo 26 April 1971 from French to "Vice Chancellor Saxon through Associate Dean Rasmussen," Estrin Papers.

[38] Ibid.

[39] Application for Certification by Board of Examiners for Clinical Engineering Certification, Estrin Papers.

[40] *Computer*, vol. 12 , no. 11, 1979.

[41] T. A. Estrin, R. Sclabassi, and R. Buchness, "Computer graphics applications to neurosurgery," *Proceedings of the First World Conference on Medical Information* (MEDINFO) (Stockholm, Sweden; 1974), pp. 831–836.

[42] T. Estrin, J. V. Wegner, and R. Bettinger, "Computer generated brain maps," *Proceedings of the San Diego Biomedical Symposium* (San Diego CA: 5–7 February 1975), pp. 369–374. The shift of attention to animal surgery was prompted by organized student opposition at UCLA to stereotaxic surgery in humans (see Estrin, "Computers, neuroscience and women . . . ").

[43] R. J. Sclabassi, R. Buchness, and T. Estrin, "Interactive graphics in the analysis of neuronal spike train data," *Computers in Biology and Medicine*, vol. 6, 1976, pp. 163–178.

[44] R. N. Harper, R. J. Sclabassi, and T. Estrin, "Time series analysis and sleep research," *IEEE Transactions on Automatic Control*, vol. AC-19, 1974, pp. 932–943, and F. K. Gregorius, T. Estrin, and P. H. Crandall, "Cervical

spondylotic radiculopathy and myelopathy," *Archives of Neurology*, vol. 33 1976, pp. 618–625.

[45] R. Buchness, T. Estrin, and J. Sue, "Use of MUMPS for interactive graphics" (abstract), *Proceedings of the 1975 MUMPS Users' Group Meeting* (Washington, D.C., 17–19 September 1975).

[46] Estrin, "Information handling in brain research using a spectrum of computers," *Proceedings of the Ninth Hawaii International Conference on Systems Sciences* (Honolulu: University of Hawaii, February 1976), pp. 68–70.

[47] Estrin, "Information handling . . . ,"pp. 68–70, and Estrin, "A low cost micro-computer system for EEG data," *Proceedings of the 29th ACEMB* (Boston MA, November 1976), p. 188.

[48] Interview 1992, p. 70.

[49] Estrin, "The UCLA Brain Research Institute. . . ."

[50] Ibid.

[51] Estrin, "Computers, neuroscience and women. . . ."

[52] T. Estrin and R. C. Uzgalis, "Information systems for patient care," *Computer*, vol. 12, no. 11, 1979, pp. 4–7.

[53] Estrin, "Health care systems," *Proceedings, The Jerusalem Conference on Information Technology* (Jerusalem: ILTAM Corporation, 1971), pp. 37–38; and Estrin, "Federal policy and health care computing," *Proceedings of the 14th Annual AAMI Conference* (Las Vegas NV, 20–24 May 1979) (abstract), p. 73.

[54] Memo 23 April 1964 from Estrin to Dean S. M. Mellinkoff, Letter 12 May 1964 from John Field to Estrin, and Letter 22 May 1973 from Estrin to Dean A. F. Rasmussen, Estrin Papers.

[55] Memo 20 December 1976 from Estrin to Deans R. O'Neill and F.A. Rasmussen, and Letter 11 April 1977 from Estrin to UC President David Saxon, Estrin Papers. See also Interview 1992, pp. 75–78.

[56] D. L. Hudson and T. Estrin, "Microcomputer-based expert system for clinical decision-making," *Proceedings of the 5th Annual Symposium on Computer Applications in Medical Care* (Washington, D.C.: IEEE Computer Society, November 1981), pp. 976–978.

[57] D. L. Hudson and T. Estrin, "Derivation of rule-based knowledge from established medical outlines," *Computers in Medicine and Biology*, vol. 14 1984, pp. 3–13.

[58] Interview 1992, pp. 94–95.

[59] Quoted in Mary Love, "A woman whose time finally came," Santa Monica CA *Evening Outlook*, 3 December 1984, p. A-6.

[60] Estrin, "Continuing engineering education at UCLA—a comprehensive approach," *Proceedings of the 1986 Conference on Continuing Engineering Education* (7–9 May 1986), pp. 439–442.

[61] Estrin wanted UCLA to become a member university of NTU and brought the head of NTU to UCLA to meet with university administrators, but no action was taken.

[62] 1981 interview of Estrin by Marshall, p. 51.

[63] Irene Carswell Peden had earlier served on the Board, but was appointed.

[64] This award is sponsored by the IEEE Aerospace and Electronic Systems Society, and preference is given to an individual who has made the contribution prior to the 37th birthday (as 36 was the age of the engineer Judith Resnick when she died in the explosion of the Challenger space shuttle).

[65] Nominations and Awards Committee; Long-Range Planning Committee; Editorial Board, *Spectrum*; Vice President, Publications Board; Editorial Board, IEEE Press; Public Information Committee; Awards Board, Edison Medal Committee; Harry Diamond Memorial Award Committee; Education Committee; USAB COMAR Committee; and History Committee. (Though not described here, some problems arose for Estrin in the course of her work for IEEE. These are discussed in the 1992 interview.)

[66] The women who had already been named IEEE Fellows were Betsy Ancker-Johnson, Jenny Rosenthal Bramley, Grace Murray Hopper, Elizabeth Laverick, and Irene Peden; Edith Clarke was an AIEE Fellow (*Washington IEEE Bulletin*, vol. 16, no. 5, 1977). Estrin was also one of the first women to attain the status of IEEE Life Member; Grace Murray Hopper attained that status earlier.

[67] Estrin, "The responsibility of engineering societies," *Proceedings of the 13th Annual AAMI Conference* (Washington, D.C., March/April 1978) (abstract), p. 87.

[68] Estrin, "Computers, neuroscience and women. . . ," Various papers relating to certification, including copies of the application and the certificate, are in the Estrin Papers.

[69] Resolution, Board of Trustees, Aerospace Corporation, December 1982, Estrin Papers.

[70] Program Description, Biotechnology Resources Program, NIH, September 1978, Estrin Papers.

[71] "Achieving our goals," talk given in 1982 to a conference of the Association of Women in Computing, Estrin Papers.

[72] An example is her failure to receive from NSF a travel grant to an international conference for which she was an invited plenary speaker and a session organizer, when younger male colleagues with smaller roles at the conference received grants. (See Interview 1992, pp. 92–93, and various letters from 1979 in the Estrin Papers.)

[73] Estrin, "Women engineers. . . . "

[74] "COMPOW goes into action," unsigned article in a 1976 IEEE public relations release, Estrin Papers.

[75] Interview 1992, p. 98.

[76] Interview 1992, p. 112.

[77] Quoted in *The Institute*, vol. 2, no. 7, 1978.

[78] The quotations are from a special section, "Women in science," in the 13 March 1992 issue of *Science*, both in articles by Ann Gibbons (the first on page 1368, the second from page 1386). The second quotation contains a quotation of Margrete Klein, director of women's programs at NSF.

[79] Ann Gibbons, "Creative solutions: electronic mentoring," *Science*, vol. 255, 1992, p. 1369.

[80] Quoted in Alice Posner, *Women in Engineering* (Skokie IL: VGM Career Horizons, 1981), p. 78.

[81] Estrin, "Women engineers. . . ."

[82] An edited excerpt from Taped Comments on Documents Sent to the Center for the History of Electrical Engineering, 1 September 1992, Estrin Papers.

[83] Letter from Estrin to Emily Sirjane 25 January 1979, Estrin Papers.

[84] Estrin, "Women engineers. . . ."

[85] Estrin, "Computers, neuroscience and women. . . ."

[86] Estrin, "Women engineers. . . ."

[87] Garry Abrams, "The desire to achieve runs deep among members of Estrin family," *Los Angeles Times*, 5 January 1986, part VI, pp. 1–3.

[88] Still other honors are the following: Distinguished Service Citation from the College of Engineering of the University of Wisconsin (1975), Outstanding Engineer of the Year Award from the California Institute for the Advancement of Engineering (1978), Distinguished Professional Contributions to Engineering Award from the National Society of Professional Engineers and California Society of Professional Engineers (1985), and Founding Fellow of the American Institute for Medical and Biological Engineering (1992).

[89] Two-page autobiographical sketch (ca. 1988) beginning "I was married when I was 17 . . . ," Estrin Papers.

CHAPTER 8

"The Magic of Your Dial"
Amos Joel
and the Advent of Electronic
Telephone Switching

AMOS JOEL

Figure 1. Amos Joel, recipient of the Kyoto Prize, member of the National Academy of Engineering, winner of the IEEE Medal of Honor, and current resident of South Orange, New Jersey.

In 1928 the Joel family of Atlantic City, New Jersey, was paid a visit by a technician from the local telephone company. Ten-year-old Amos, the Joels' only son, watched as the repairman installed a new telephone in the living room. He recalled later that "at that time we had a desk-stand telephone, so they replaced it with one with a dial, and after a certain day you were supposed to start using the dial. I got very curious about this. I said to myself, 'well, how does this thing work?' I wrote a letter to the New Jersey Bell and got a booklet back—*The Magic of Your Dial*. But it didn't tell me enough to satisfy me."[1]

That precocious curiosity persisted, leading Joel to a career in telephone research and engineering. By the time he retired in 1983, he was a leading international figure in one of the most important areas of telecommunications: electronic telephone switching. This new technology emerged from Bell Telephone Laboratories, the corporate research and development facilities of American Telephone and Telegraph, where Joel spent his entire professional career. At Bell Labs, engineers inspired by the promise of electronics guided the development of new electronic switching networks from simple prototypes to computerized central offices of massive scale. Joel rose from a fabricator of electromechanical switchgear in the late 1930s to become one of the leading architects and designers of AT&T's first regular-production electronic switching network in the 1960s.

A prolific inventor with over seventy patents, he contributed to the evolution of electronic switching at each stage of its development. Throughout his career, he also promoted the study of switching as a formal field of engineering science in his technical articles, teaching, and volunteer activities for his professional society. Later in his career, he became an accomplished historian; his publications include a comprehensive history of switching in the Bell System and a more recent book that he coauthored on the international development of electronic switching. This chapter examines Joel's career and concentrates on the process by which electronic switching technologies developed at Bell Labs.

Joel's Early Years

Amos Joel, Jr., was born in Philadelphia on 12 March 1918. During his childhood, his family moved several times, and his earliest memories date from the period after his father moved the family to the seaside town of Atlantic City, New Jersey. His father worked as a traveling salesman for a clothing manufacturer in Philadelphia and was frequently on the road. In 1929 he took a position working for a relative at the famous New York City clothier, the A. Sulka Company, where he eventually became an executive vice president. Young Amos attended public schools in Atlantic City and later at the famous DeWitt Clinton High School in the Bronx when the family moved to New York City.

Amos became interested in electromechanical technologies early in life. As a youth, he played with electric trains and an Erector Set, and experimented with a crystal radio. Model railroads became a consuming hobby, and he constructed elaborate track layouts and signaling circuits. In his teens, he also played the saxophone and clarinet.

His early curiosity about the new dial telephone began at the family home, but it did not stop there. His imagination followed the wires out to the telephone pole and down the streets of Atlantic City, straight to the central switching office. At age ten, he began to pursue his engineering career by querying the New Jersey Bell Company for information about how the dial telephone system worked. He was disappointed that they did not invite him to visit the telephone exchange, but his interest continued undiminished. In 1929 his family moved to an apartment on West 86th Street in Manhattan, and there he set up a private telephone system to connect him to his neighborhood friends. At each end of his block, Joel built a crude manual switchboard from knife switches. He strung wires along the fences in back of the 86th Street high-rises and into his friends' apartments.

The long arm of the telephone company inevitably caught up with Joel. While attending the switchboard one day, the telephone rang. A voice, louder and clearer than was usual on his network, announced "this is the telephone man. Where are you located?" The Depression had left many apartments empty in New York, although they were still equipped with leased telephones. These telephones, "liberated" by Joel, had been put to work in his private network. According to Joel, New York Telephone did not take kindly to his private service, but he maintained that "it was fun building it."[2]

With an ever-growing interest in the telephone system, Joel wrote at age thirteen to American Telephone and Telegraph in regard to one of their newly developed switching systems. An employee referred him to the relevant patent literature. Thus Joel began another hobby: collecting patents and technical information about telephone switching. At the local library he discovered the monthly journal of patent announcements, the Patent Office *Official Gazette*, and began to order copies of every switching patent he could identify.

Joel's knowledge of switching systems grew, and in 1931 he decided to design his own system. The product, which he called the Joel All-Relay Dial System, utilized small, general-purpose electromechanical relays. When he started to build his prototype, however, he learned an important lesson: the cost of relays— about a dollar each—made the system prohibitively expensive.

There was one opportunity I had a little later, just before I went to college, which was in 1936. I had acquired a copy of a little pamphlet put out by the New York Telephone Company for their people. Part of a training course or something. It was a beautifully done pamphlet about the panel dial system, which was one of the major dial systems at the time. It was written by a man by the name of Ellsworth H. Goldsmith. I remember that. I got curious and wrote him a letter. He invited me down to the Telephone Company, and I met him. He didn't take me through the telephone office or anything, but I had a chance to talk to him about it and learn more. I learned that he was a horologist, interested in clocks. He put clocks together. That was his hobby. But he happened to have a job in the Telephone Company writing course material.

That was the only time I ever had any contact with people from the Telephone Company other than writing to them and trying to get information, which, I would say, on the whole, was not very satisfactory. In fact, it led me later to try to do something about this. Later on in my career I felt that switching got the short end when it came to telecommunications. Everybody talked about transmission, and they did a great job of teaching it in colleges and schools. Transmission had been reduced to at least a quasi-science so that you could express mathematically the various relationships that take place in transmission. But you couldn't do this in switching. And I wondered why. Why couldn't we do a better job of it? I felt that part of this was not enough dissemination of what goes on in the business of switching.[3]

Joel at MIT: The Making of an Engineer

In 1933, as Joel considered high-school graduation, he began to plan for college. He had his sights set firmly on becoming an electrical engineer. He wanted to go to the Massachusetts Institute of Technology (MIT), but worried that his parents would not be able to afford it. His parents did, however, manage to find the resources, and he matriculated in the autumn following his high school graduation in 1936.

At MIT Joel found that the electrical engineering department was in the process of instituting changes to the curriculum, using a new series of textbooks to shift part of the instruction from power engineering to some of the leading-edge electronics fields, especially communications. He met a professor, Carleton Tucker, who was interested in telephone transmission and who gave him encouragement. Joel even got the chance to prove his facility with switching systems by repairing the electrical engineering department's telephone equipment. Through Tucker, Joel made his first contacts in the telephone industry and gained hands-on exposure to a variety of telephone equipment.[4]

Electromagnetic theory, important in telephone communications, was part of the engineering curriculum at MIT, but other communications-related topics like switching were not offered. Many electrical engineering courses were concerned with electrical power, and lectures on the telephone system concentrated on the transmission of signals. One reason may have been that there was no real theoretical basis for switching, a problem that Joel would address later in his career.

Joel made many friends among the faculty and students. One acquaintance he remembers particularly well was Claude Shannon. They frequently walked the streets of Boston (and later Manhattan, when they both worked at Bell Laboratories) by night, talking "shop." Shannon recalls that these talks with Joel were the inspiration for the subject of his master's thesis, the application of Boolean alegebra to the design of relay and switch circuits. Shannon went on to make fundamental contributions to the information theory that underlies much subsequent research on switching theory and work with digital circuit design.[5]

Joel earned money in college by working in the office at one of the dormitories. To his delight, his duties included running the dorm switchboard. He was on this job one day when he met his future wife, a student at a nearby women's college who came to MIT with a friend. Joel asked her for a date on the spot and, in his words, "one thing led to another."[6] Joel took his date to his dorm room, where he attempted to explain some of the switching patents that he had posted on his wall. "She went home and told her father I was crazy," Joel recalls, but that did not stop her from eventually marrying him.[7]

As a senior, Joel began thinking seriously about employment after graduation. He had long since set his sights on Bell Telephone Laboratories. But with the lingering Depression, the job market was extremely tight, and Joel's first letters to Bell did not receive a positive response. His professors tried unsuccessfully to pull strings for him. He finally put aside his highest hope and contacted a smaller manufacturer, the Automatic Electric Company.

Automatic Electric, which manufactured telephone equipment in competition with AT&T's Western Electric Company, brought Joel to its Chicago facility for a job interview. Unbeknownst to him, things were finally stirring back at Bell Labs. One of Bell's patent attorneys had recommended Joel at the insistence of his son, who was Joel's classmate at MIT. Toward the end of a day of interviewing in Chicago, Joel received a telephone call from the manager of the local A. Sulka store. He was told not to accept any offer from Automatic Electric until he had spoken privately to him. Joel made haste to the store, where he received the unexpected good news that Bell Labs now wanted to speak to him.

Bell Labs had finally noticed Joel's unusual qualifications and experience and contacted Carleton Tucker at MIT. Tucker informed Bell about Joel's trip to Automatic Electric in Chicago. This was enough to spur Bell's eleventh-hour call for an interview. Joel was overjoyed with the opportunity and went immediately from Chicago to New York for an interview at the Labs. Both companies offered him jobs, but Automatic Electric initially offered a higher salary. When Bell Labs matched the offer, Joel gladly accepted.

Early Years at Bell Labs: The Switching Art in Transition

In July 1940 Joel reported for work at Bell Laboratories on West Street in New York City. At that time there was only one major Bell Labs site, though during the next few years the new labs opened in Murray Hill, New Jersey. Formed in 1925, Bell Labs had become one of the largest and most prestigious private research and development organizations in the world by 1940. The organization had a hierarchical structure of management that followed the model of American manufacturing companies. The Labs employed a large number of engineers and scientists, who might be assigned to work on individual projects, direct those projects, or manage large groups of projects referred to as "laboratories."[8]

Despite the impressive knowledge of switching he brought with him to Bell Labs, Joel was assigned to the same training program as the other new engineers, which involved being moved from department to department to get a feel for the entire operation. His first assignment was in a shop that assembled equipment. The work was tedious, and after a few days he pleaded with his supervisor for reassignment. After another week, he was moved to the relay section, where he built and adjusted special-purpose relays. This work proved to be more rewarding, although within a short time he was relieved to be sent to another lab to work on the design of relays. He discovered that one of his new bosses, F. J. Scudder, was someone whose work he already knew well from his patent collection. He soon met many of the other Bell Labs engineers named in telephone switching patents.[9]

Bell Laboratories played a major role during World War II in the development of electronics and communications technologies for the military, and during this period Joel cut his teeth in the fields of electronics and computers. The Labs' contributions to the war actually began in a very modest way in the late 1930s with a small radar project. Radar research expanded greatly in the early 1940s, but by then it was only one of many military projects at Bell Labs. The number of staff members working under military contract grew explosively from 200 in 1940 to over 2000 (of the 2500 employees) in 1943. Bell research eventually ranged over a whole landscape of electrical and electronics technologies, from radar and sonar, to improved radio and wire communications, to proximity fuzes and fire-control computers.[10]

Shortly after the United States declared war in 1941, developmental work on telephone switching and other civilian telephone technologies virtually ceased. Joel joined a project working on military cryptographic systems that used electromechanical relays and complex switching circuits. He recalled his relish at working on something "really brand new."[11] His first assignment was the development of a machine to transmit and receive coded teletype messages. Standard teletype machines transmitted information in the form of pulses, and the Bell Labs' design made these pulses unintelligible to any receiving station not equipped with the proper decoding equipment. Early in the war, the Army Signal Corps and the Navy adopted this technology for secret communications.[12]

Joel also participated in a project sponsored by the National Research Council to develop a voice encryption system, so that secure voice transmissions could be made by wire from the battlefield. The resulting device filtered out several narrow bands of frequencies from the signal and transmitted them in jumbled order. Only a specially equipped receiver could reorder the encrypted message properly. Since such machines were also well-known to the Germans, the coding could generally be broken after only a few hours of expert analysis. For this reason, these machines were used only for applications such as relaying battlefield instructions, where the secrecy of the transmissions had value for only a short time.

Joel's group later became involved in an effort to digitize speech for the top secret scrambled-signal voice transmission system known as Project X. Bell Labs researchers in the 1930s had already developed the Vocoder, a device for splitting speech signals into a multitude of different frequency bands. Now Bell designed electronic circuits to digitize these discrete channels and combine them with other, randomly generated signals. As in the earlier Teletype devices, receiving and transmitting equipment contained matching versions of the code used to encrypt and decipher the signal.[13]

The original version of the Project X system used phonograph recordings of the decoding information, with just two copies pressed before the master was destroyed. Joel and other researchers designed a relay-based version that, while less secure than the phonograph type, could be used to set up and test the equipment (saving the records for actual operations) or in installations requiring less stringent security.

Later Joel worked on another enciphering system that employed a technique called "pulse code modulation." He received one of his first patents for this work, a solution to the crucial problem of synchronizing the digital data at each end of the system. This project gave him an opportunity to become much more familiar with electronics, a subject that he had studied only briefly at MIT but which played an important role in his later career.[14]

In January 1940 Bell Labs placed into operation an electromechanical complex number calculator built from ordinary telephone relays by two of its engineers, George Stibitz and Samuel Williams. It was used for multiplying and dividing complex numbers in connection with research and development.[15] Because of this experience, Bell Labs was contracted by the government to build a series of electromechanical calculating machines for the war effort. Three such machines—the Relay Interpolator, the Model III, and the Model IV—were built to aid in the design and simulation of automatic aiming devices for antiaircraft guns. Near the end of the war, Bell Labs was contracted to build a much larger and more complex relay calculator, the Model V, for use in producing artillery firing tables. The first one was delivered to the Army in 1946. Rather than having arithmetic circuits for doing calculations, it made extensive use of look-up on hardwired addition tables.[16]

Joel was assigned to work on the Model V computer. He was given the task of designing the most complicated circuit, which had to do with trigonometric functions, logarithms, and tape block searching—a very unusual circuit that contained its own memory. Because of this experience, once the war was over Joel was assigned to designing computing equipment for the Automatic Message Accounting system being developed for implementation with the first Crossbar System in Philadelphia. Joel's wartime experience in digital electronics and computing was extremely valuable to his postwar work on electronic telephone switching.

Joel and Electronics: The Automation of Long Distance Dialing

Joel's first assignment after the war returned him to the design of switchgear.

Bell Labs engineers and managers were enthusiastic about the possibilities of the new technologies developed during the war, including computers and improved relays and vacuum tubes. Further, within a few years the Labs announced the invention of the transistor, and speculation ran high regarding the opportunities presented by electronics. Switching was one area in which Bell pursued those opportunities, resulting in rapid technological innovation during the three decades following 1945. During these years Joel advanced from a designer in large projects to a systems engineer, and thus had an increasingly important role in both the promotion of new switching technologies and in their realization. He eventually became instrumental in a revolution in telephone technology, the introduction of electronic switching. Its revolutionary aspects are evident in the changes occurring in the telephone industry today. As older equipment has been replaced by electronic switchgear that is computer-controlled, the range of services, flexibility, reliability, and speed of the telephone network has vastly improved.

For AT&T the road to the new technology was marked by experimentation, false starts, and temporary reversals—just as in most technological innovation. The company experimented with a number of transitional technologies before attempting a fully electronic system. Joel contributed at every step of the way, either in the design of particular machines or in the management of projects leading to new technological systems. One of the first large postwar projects to which Joel was assigned resulted in his first major computer project, the Automatic Message Accounting Computer.

AT&T had long been interested in automating the routing of long-distance telephone calls as much as possible, and by the late 1940s the drive to automate steered research toward electronics. The introduction of the dial telephone in the 1920s was matched by innovations in electromechanical switching machines that could operate as automatic switchboards. The first generation of automatic switching technology, called the Step-By-Step System, consisted of a machine with sectors of electrical contacts and a small number of moving contacts. These contacts interconnected incoming and outgoing lines, which converged at central offices. When a customer dialed a number, the telephone set sent electrical pulses to the switchgear at the central office. Those pulses controlled the movement of the switchgear directly, causing it to connect the incoming line with the appropriate outgoing set of contacts. Each number that was dialed acted upon a particular set of switches, establishing a complete connection through a series of steps. This system worked well for small, self-contained communities, but it could not be easily or economically expanded to serve large cities, nor could it be modified for automatic long-distance use.

In the 1920s and 1930s, two new generations of switching technology emerged to address these problems. The first was the Panel System, so named because its automatic components moved up and down along contacts mounted on a long panel. Panel machinery, which was used experimentally as early as 1915, was built by Western Electric until 1950 and used until 1982. (See Figure 2.) A second type of electromechanical switch was the Crossbar System, first used in 1938. Unlike the previous two generations of equipment, the Crossbar relied on a matrix of contacts, rather than contacts that moved through large distances. The Crossbar could handle more connections per machine and was designed to make connections faster and more quietly.[17] With the Panel and Crossbar systems, Bell engineers also developed the concept of "Common Control," a method for controlling the system that employed various ancillary devices to store dial pulses temporarily and make more effective decisions on call routing. Common Control became the central focus in switching design, resulting by the 1950s in highly complex and specialized machines such as markers, registers, and senders.[18] The complexity of these machines increased with the size of the network, creating problems with cost, reliability, speed of operation, and physical size. All of these problems pointed the Labs toward the development of new component technologies. However, the concept of Common Control carried over into the electronic era, becoming the foundation of specialized switching computers.[19]

The need for even faster, more sophisticated switching equipment was more keenly felt in the postwar period when AT&T established the goal of nationwide direct long-distance dialing. Toll calls previously had passed through central long-distance exchanges, where human operators manually established connections to exchanges near the call destination and wrote charge tickets for billing. The first step, the institution of a national, uniform numbering system with area codes (so that geographic areas could be accessed by automatic switches), was followed by the installation of automatic equipment to make direct long-distance dialing technically feasible. This, in turn, required expensive modifications to many local exchanges, so that the switching systems could deal with the extra dialed digits representing area code numbers.[20]

The new switching system, or modifications to older ones that allowed direct dialing, was accompanied by an automatic machine to record information about long-distance calls for billing purposes. During the war, a Bell researcher named W. W. Carpenter had proposed a system for automatic message accounting (AMA) that could record information about long-distance calls in a machine-readable form.[21] Carpenter's AMA recorder punched on a paper tape a coded message providing a record of the origin, destination, and elapsed time of telephone calls. Joel was assigned the task of designing a complementary accounting computer to process this information so that callers could be billed.

The AMA computer resulted in a very lengthy US patent with over 250 claims. Over 100 of these computers were installed in centralized AMA installations, called accounting centers, dispersed across the country. The punched-tape output of the AMA machines was eventually converted to magnetic tape, and the accounting computers went out of service. The electromechanical AMA system was a useful technology for a time and contributed significantly to the ongoing drive toward full long-distance automation.[22]

Figure 2.   Photograph of a panel switching frame.

When in the late 1940s it became a goal at Bell to apply electronics to switching, three basic lines of inquiry emerged. The first was to replace the mechanical relays used in the switching machines with electronic relays, which were thought to be faster, more reliable, less bulky, and perhaps less expensive. The second, and more challenging, goal was to design electronic

Common Control devices that would replace electromechanical equivalents and increase the efficiency and flexibility of the system. Third, the computer technology under development at Bell Labs and in research institutions around the world seemed to be the key to a fully electronic network. Engineers began to adapt computer technology to suit the requirements of telephone switching.

After the war Bell Labs expanded its investigations of the application of electronics to switching. Vacuum tubes had been brought to a high level of development at Bell in the 1920s and were already used widely in long-distance amplifiers. In the early 1940s Bell researchers, experimenting with gas-filled diode tubes, began to appreciate the increases in switching speed possible by employing electronics in place of mechanical relays. Because of the mass of their moving contacts, relays had inherent physical limitations on their maximum speed.[23] New switching devices demanded high-speed electronics in order to be useful. Interest in electronics grew after the war, as many Bell Labs employees returned to civilian work with experience in military radar and communications electronics. During the 1940s and 1950s, Joel and his colleagues found that electronics was the key to improved Common Control and switching devices. Other technologies developed for computers, including magnetic drum, cathode-ray tube, and ferrite memory, were adopted by Bell Labs researchers for use in a range of switching machines.[24]

In 1951 C. E. Brooks, an engineer at Bell's West Street laboratories in New York City, proposed an electronic replacement for Panel and Step-by-Step central offices. The new system was to consist of three main sections: first, the switching network and control circuits; second, an accounting facility, which was envisioned as being located some distance away from the central office; and third, a Remote Line Concentrator, a small automatic switching machine similar in function to the a Private Branch Exchange (PBX).[25]

Whereas PBXs had usually been marketed to companies in office buildings, the Line Concentrator was intended for somewhat smaller buildings, such as apartment complexes. It was usually installed on the outside of a building in contrast to the PBX, which was mounted in a special room, often in a basement. In addition to filling a special market niche, the Line Concentrator was also a test case for the use of new switching technologies and physical design techniques. The high cost of maintenance, a chronic problem for switching equipment installed on a customer's premises, was addressed in part through the placement of critical circuits on plug-in circuit boards, which could easily be replaced when circuits failed.[26]

The function of the Line Concentrator was to allow a large number of telephones to be connected to a central office over a much smaller number of trunk lines. Taking into account the calling characteristics of its customers, AT&T determined the number of trunk lines based upon the number of telephones in use and the average volume of usage. This increased the efficiency of the network and eliminated some of the cost of stringing new lines. In prototype form, the Line Concentrators served sixty customers with only ten lines to the central office.

The switches in the Concentrator used a special switching device consisting of a reed relay and gas tube, several of which were built into a plug-in module. Telephone companies installed concentrators in LaGrange, Illinois, Englewood, New Jersey, and Freeport, Long Island, and the system proved to be reliable.[27]

In addition to the equipment at the customer's premises, special electronic Common Control machines with magnetic drum memories were used at the central office in conjunction with the Concentrator. Though not invented at AT&T, magnetic drums were incorporated into a series of early Bell Labs designs. Despite its technical success, the Remote Line Concentrator was not implemented in the Bell System. According to Joel, the additional development was needed to guarantee its reliability in long-term service. AT&T also decided that the small potential market for the Concentrator did not justify the additional development costs that would have been incurred in preparing the design for production. The Line Concentrator nonetheless demonstrated the possibilities of the gas-tube reed relay, which was used in later switching equipment.[28]

JOEL: . . . At the same time that Chet Brooks sold this originally, the idea of exploring electronic switching when the transistor was first invented, there were other people at high levels in Bell Labs who wrote similar memos: Now is the time we ought to be doing something about switching . . . getting away from electromechanical. So there were other people thinking along that line. But Chet was the one who was pushing so hard that he wanted to set up a group to do the whole thing. The other people were just writing memos about it.

ASPRAY: What were the limitations on transistors at the time? They weren't reliable, you couldn't manufacture them and attain stable characteristics.

JOEL: They were awful! In fact, one of the things that came out of this first two years of our exploratory work was [the realization that the new technology] was so poor, so unreliable, so unpredictable, that we should get some smaller project moving first. Indeed, in my group we started two different projects that were only tangentially related to a completely electronic switching system. [One project was an] electronic remote concentrator that we could apply to [the existing] electromechanical systems.

ASPRAY: What is a remote concentrator?

JOEL: *The idea is that with an apartment building or a housing development, instead of running a pair of wires from every apartment or house to the central office and there picking up the first point in the switch, have the first point in the switch out there. . . . It had been a dream of switching people for years, but they had no technology that would do it well. . . . Here we had transistors that we thought could do a lot better. We [built the transistorized remote concentrator and] actually put three of them in service.*

ASPRAY: *But this never became a widely-distributed technology?*

JOEL: *No, not for a long time. In fact, we did so much to prove the idea that remote concentrators were good that the electromechanical people went out and designed an electromechanical remote concentrator that was a lot cheaper than the electronic one. They sold their design, and it was placed into production and service.*[29]

Joel contributed to a series of other transitional switching devices employing electronics technology during the 1950s. Implementing direct-dialing of long distance numbers required significant changes in the various switching networks in use around the country. Beyond the standardizing of area codes, Bell Labs engineers had to find ways for Step-By-Step, Panel, and Crossbar switches then in use to work in the new system, which supplanted 7-digit telephone numbers with 10-digit numbers (7 digits plus an area code) and long-distance access codes. Even in the Step-By-Step system, where dial pulses controlled the switching directly, Bell engineers felt that something resembling Common Control would have to be added to the machines. The space- and time-saving possibilities offered by electronic components presented opportunities to make compact new devices that could add more automatic controls for switching machines, and Joel conducted exploratory development in this area.

The introduction of Direct Distance Dialing (DDD) necessitated the use of an access code, originally 211 but later just 1, to initiate a long-distance call. Joel's group worked on a "translator," a kind of automatic control device to adapt existing switching systems to the new long-distance dialing procedure. They devised a machine called the Magnetic Drum Auxiliary Sender (MDAS), which recorded incoming digits, made decisions about where to send them, and replayed the recording into the proper lines. One novel feature of MDAS was its ability to scan electronically the telephone lines for incoming information. When the device detected an incoming call, the dialed numbers were stored as magnetized regions on the magnetic drum memory while the Common Control made decisions about call routing.

The MDAS replaced a large number of conventional registers (electro-mechanical common-control devices) by using magnetic memory. The pulse information constituting a dialed number could be stored on about a square inch of the drum surface. The ability to store a number in this way eliminated about fifty relays and several cubic feet of space in an electromechanical register. The MDAS was a hybrid electronic device representing the transition from vacuum tube to semiconductor electronics, employing logic circuits built with transistors and semiconductor diodes but with read-write circuits that used conventional vacuum tubes.[30]

Joel carried the development of the MDAS through to the point of drawing up detailed plans for production. When in 1954 AT&T determined that the design would not be economically feasible, development ceased. One economic issue that doomed the MDAS was its potential unreliability in service, expressed in terms of "maintainability," reflecting AT&T's traditional high standards of engineering. The vacuum tubes and roller bearings used in the drum assembly had maximum expected life spans of only one or two years, which the company deemed unacceptable. Yet the MDAS proved valuable as a demonstration of the use of magnetic recording in telephone switching, which reappeared in later systems.[31]

Despite these abortive efforts, AT&T remained committed to implementing electronic equipment in the telephone network. The application of semiconductors in electronic switching equipment at Bell Labs began shortly after the invention of the transistor and took place alongside work with vacuum tube electronics. Joel's first exposure to transistors came as he was finishing up the AMA computer in 1947. In that year there came a "fateful day," when "they got ... all the members of the technical staff that they could fit in . . . the auditorium and told us about the transistor. And then they sent us back and said, 'What can you do with it?' Then I started thinking, boy, this is really the opportunity."[32]

One of the central problems in lowering the cost of switching networks, and thus making it more feasible to construct larger ones, was the cost of relays and related components. Joel had learned this at an early age, when his "Joel All Relay Dial System" proved too expensive to build. Another factor was speed: A faster switching device could obviously be more productive in a given amount of time than a slower device. A third issue was reliability. One approach to the high cost of switching equipment was to make it reliable enough to operate for decades without requiring replacement. Bell researchers hoped that similar standards of reliability could be built into semiconductor-based machines.

Finally, perhaps the most important advantage of the transistor was its low power and heat dissipation. Bell Labs engineers perceived a substantial savings in operating costs to be possible with compact transistorized switching networks. As semiconductor-based switching came on line in the 1970s and 1980s, Bell System Companies greatly reduced the number of its leased buildings in many cities, which reflected the smaller size of the new systems.

Semiconductor Switching in the Public Network:
Joel and the Making of the ESS

Bell Labs research in semiconductor switching dated to the 1930s, when researchers experimented with rectifiers made from copper oxide and other metals. By the mid 1940s, projects were underway to use germanium to construct a solid-state amplifier. When the transistor emerged in 1947, it was immediately apparent that it had uses as both an amplifier and a switch. Although the first transistors were unreliable and expensive, they captured an influential cadre of supporters. The weight of over a decade of research and the enthusiasm of certain key engineers and administrators overcame the reluctance of some designers, particularly those in switching, to adopt this new technology.[33]

When Bell Labs began to produce experimental transistors in quantity after 1948, Joel (having moved from product development to systems engineering) was asked to join a "browser" group on transistor applications. Bell Labs regularly assembled browser groups to investigate opportunities presented by new technologies and suggest long-range strategies for the company. The ideas suggested by Joel's browser group were predictions rather than practical design proposals. But to Joel and other researchers these investigations served to reveal the wide scope of possibilities for the transistor in switching. Joel joined a growing party of Bell Labs researchers enamored of these possibilities. In the early 1950s, these researchers went to great lengths to lobby for transistor development projects. In the case of electronic switching, the support of upper management helped ensure that the projects continued to be funded over a long period of product development.[34]

In the early 1950s, the ascendancy of semiconductor electronics in the telephone system was not at all apparent. The transistorized-switching exploratory group, of which Joel became a member, was not the only group "browsing" new switching ideas. Other researchers were simultaneously developing ideas such as gas-tube switching and improved electromechanical relays. Additional competition came from the existing development groups, inasmuch as many engineers in the switching field opposed electronics research because it encroached on the electromechanical art, which they believed was superior in reliability and cost. When Joel's group began to campaign for large sums of exploratory development money, opposition from these other groups mounted. In response, a member of Bell Labs' Systems Engineering section and an ardent supporter of electronics, Chester Brooks, drafted a long and apparently influential memo promoting the application of transistors in switching and forecasting a great future for them in the Bell system.[35]

With support at high levels within AT&T, including that of Bell Labs President Mervin J. Kelly, the labs established a new Exploratory Department in 1952, headed by Joel's associate Bill Keister. Joel was chosen to supervise the architectural design of a new transistorized switching system. This included the Remote Line Concentrator and the MDAS. The projects were moved from New York to the laboratories in Whippany, New Jersey, some thirty miles away. About thirty people were brought in to work on the project, with Joel as one of four supervisors. The group ran into problems early on with Chester Brooks, whose visionary ideas had only scant grounding in knowledge of electronics. The publicity that would accompany the opening of the new office made the choice of its location political. Brooks chose Morris, Illinois, a small city in the jurisdiction of the Illinois Bell company where he had friends. At his insistence, the mission of the switching project was changed from an experimental investigation to the construction of a working prototype, to be called the Electronic Central Office (ECO).

When the basic architecture was laid out, Bell Labs brought in engineers from Western Electric to assist in the construction of the first ECO. Because Western Electric was to be AT&T's manufacturer of commercial electronic switchgear, the Labs began working with them on the system prototype. This ensured that the final form of the new switching system would be designed with production in mind.[36]

Several months of work resulted in the group's earliest transistorized switching devices. They were "so poor, so unreliable, so unpredictable," that the group began revising earlier deadlines for the Morris project.[37] In the meantime, the work of other Bell Labs researchers was steadily improving the reliability of transistors. By 1953 the company decided to fund the construction of laboratory models of an electronic central office, with services comparable to a Panel- or Crossbar-equipped central office plus Remote Line Concentrators. The work was assigned to the department at the Whippany Lab, where Joel had now been promoted to head.[38]

As the ECO laboratory model emerged, it came to be known as the "pre-Morris" system, as distinct from the much more elaborate installation planned for Morris, Illinois. The model would employ an electronic device to scan telephone lines for originating calls and a central control unit capable of storing call information and sending out commands to the distribution network. The network unit was made from a matrix of gas tubes that formed the electrical connections between telephone lines accessed through remote line concentrators. It was modularly constructed, consisting of plug-in circuit boards holding computer-style "building block" circuits such as logic gates. The system also incorporated stored-program control utilizing a cathode-ray tube memory, instead of the older method of controlling the machines' functions with hard-wired circuits.[39]

While Joel worked on the system plan as a whole, other designers engineered the memories, the gas tube switching network, and the trunk circuits. Initial studies recommended borrowing from digital computer technology either the magnetic drum or the Williams cathode-ray tube (CRT). Bell Labs designers instead developed a new type of cathode-ray tube memory. The Williams tube employed an internal phosphor (which glowed when bombarded with electrons) and an external capacitance-coupled plate. Bell Labs researchers placed a capacitance plate inside the tube, resulting in greater speed, capacity, and reliability. This device, named the Barrier Grid Store, was used as short-term memory for storing call data.

For longer-term memory, Ray Ketchledge, leader of the subsystems design team, proposed another type of CRT memory, the Flying Spot Store (FSS). In this tube, the electron beam struck a phosphor-coated plate, causing it to give off light at each storage location. (See Figure 3.) The light then passed through a special photographic plate holding information in the form of opaque or transparent regions that passed or blocked the light. Light passing through the plate was detected by photomultiplier tubes. Lenses inside the FSS could direct the light to several photographic plates simultaneously, making it possible for the device to manipulate words of 32 bits in parallel.

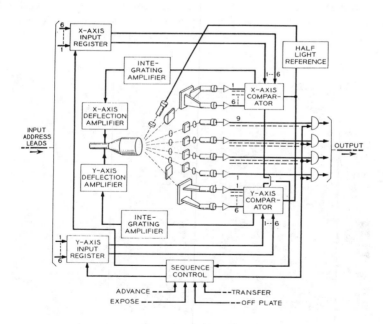

Figure 3. Simplified schematic of the Flying Spot Store.

Initially the FSS was intended to store semi-permanent information such as the directory number associated with a switching network terminal. As Ketchledge continued his work, an expanded capability of the FSS became apparent. By adding servo control to the cathode ray beam location, he was able to add considerably more storage locations (words) and increase reliability. The photographic plates used to store information were made using the FSS itself.[40]

Work on the Common Control circuits initially proved troublesome. The group started out with the goal of completely eliminating relays but discovered that the logic circuits required to do so were inordinately complex. One engineer in Joel's department, W. A. Budlong, suggested the use of a stored program, an idea from the computer field that seemed possible given the increased capability of the flying spot store. The architecture that the group devised was unusual for its day. It was designed to allow the switching equipment to deal with a very large number of inputs (incoming telephone calls) simultaneously.[41] This real-time design demanded a large amount of fast memory and high-speed logic. The size of the planned switching system was such that it could handle at least 50,000 individual lines (more than twice the capacity of a standard Crossbar system), and the engineers expected about 50,000 simultaneous call attempts during busy hours. In addition to his work on architecture, Joel was the major contributor to the overall system development and design coordination for this project, which was demonstrated in 1955.[42] (See Figure 4.)

Figure 4. The "pre-Morris" system: a laboratory model electronic switching system that incorporated stored-program control.

The actual Morris system was a greatly enlarged version of the pre-Morris laboratory apparatus, reflecting the rapid pace of change in computer and electronics technologies as well as the lessons learned in the building of the pre-Morris system. The switching network used a special cold-cathode neon tube as the "crosspoint" or connecting element, rather than the anticipated transistors, because transistors had proven incompatible with the high currents used in the existing network. The crosspoint tubes operated both as switches and amplifiers. The normal 90-volt signal used to ring a subscriber telephone could not be passed through the gas tubes, so the company designed special equipment for this purpose, including a device to send an audio frequency ringing signal to the subscriber's telephone and a special telephone with a transistor amplifier and miniature loudspeaker. Other components in the system made extensive use of semiconductor diodes and transistors.

The Morris system used a much larger and faster cathode-ray (also known as the "flying spot") memory, providing 2.2 million bits of storage. A new comparator-based feedback system substantially increased the speed and accuracy of the memory tube. The electron beam in the memory tube se-quentially scanned a developed photograph plate, which represented the program and other semipermanent information requirements, passing through it and striking photodetectors on the far side. Another CRT memory, called the Barrier Grid Store, was used as an electronic "scratchpad" to retain briefly instructions and telephone number information before dialing telephone numbers.[43]

The program used at Morris had 50,000 words of 25 bits each. This was larger than many computers of the day, but somewhat smaller than the other major real-time computer application of the era, the SAGE air defense system developed for the Air Force, which had a program of 75,000 words.[44] A third of the program was devoted to telephone operation, while the remainder was used for maintenance and administration functions.

The Morris equipment could, for example, self-diagnose equipment failures. Previous generations of electromechanical equipment had addressed the problems of fault diagnosis and maintenance, two major operating expenses in switchgear. Since the introduction of Common Control, it had been possible for electromechanical switchgear to automatically identify technical problems and indicate or otherwise make a record of those problems for the purposes of maintenance. As much more sophisticated electronic equipment began to be used, maintenance problems grew. The new electronic system carried self-diagnosis to a new level of sophistication, obviating much of the skill necessary to maintain electromechanical switches. At Morris, the task of the maintenance personnel was only that of looking up a printed trouble report in a maintenance dictionary to find out the location of a defective plug-in package in order to replace it.[45] (See Figure 5.)

During the development of this system, the group had to defend and justify its work continuously. One of the major selling points of the programmed system was that it would make future changes in the system as easy as changing the program. But, as Joel recalls, "We didn't tell them how hard it was going to be to change the memory. And how many people it took to write the program or anything like that." In fact, he says, "we didn't know [that] ourselves at that time."[46]

In March 1958 the design groups visited the pre-Morris installation for the inaugural tests of the system. A series of carefully determined tests was planned to check each stage in the system before it underwent further system tests. Casting all this planning aside, one of the supervisors in charge of the system tests instead turned to Joel and said, "Well, dial! Dial! Dial the number and see if the call goes through!" It did, and the era of electronic, stored-program switching began.[47]

In June 1960 Bell Labs tested the Morris facility by placing it in service on the public network. The cost of the Morris installation was much higher than originally anticipated. When the experimental version had been completed, the project leaders had asked for an additional $10 million from AT&T. By the time the first lines had been "cut over" in Morris, the cost had exploded to over $100 million. Based upon these demonstrations and the project management's predictions of the potential for service improvement, increased revenue, and reduced operating expenses, AT&T continued to fund the stored program electronic switching effort even though its development costs had increased tenfold.

At its peak, the Morris system serviced 434 telephones, but it remained in service only until early 1962. The new switching system proved highly reliable, but almost all of its major components were obsolete even before it went into service.[48] This was because Bell Labs and other institutions continued to develop new component technologies at a rapid pace in the late 1950s and early 1960s. One was the twistor memory,[49] an inexpensive semi-permanent magnetic memory component made of a permalloy tape wrapped in a spiral around a conductor which replaced the FSS. Another technology addressed the problem of integrating voltage-sensitive electronics into the rest of the existing telephone network, which used a 90-volt ring signal. When semiconductor switching research began in the early 1950s, a transistor capable of withstanding high voltages was thought to be within reach. But at the end of the decade a cost-effective transistor switch was still not available. In response, the Ferreed switch,[50] a nonelectronic system, was developed for the electronic switching system, ESS, described below. However, semiconductor diodes and junction transistors were used extensively in the low-voltage circuits in control devices, such as the machines used to scan lines and detect dialing pulses.[51] The next generation of electronic switchgear control retained the concept of a stored program, which had proven highly efficient, but used the new magnetic twistor hardware in place of the Flying Spot Store. The twistor used cards with magnetic spots to store semipermanent information such as the program and translations.

Figure 5. The concentration and distribution networks of the Morris system.

Development of the model Number 1 Electronic Switching System (ESS), the first commercial version of the electronic central office, began in 1958 just after the pre-Morris system had been successfully operated and new technology had been chosen for the commercial system. Joel was in charge of the architecture of this new system. The emphasis was not only on the new technology, but also the cost of the system compared with the electromechanical systems it was to replace. In 1962 Western Electric began to manufacture the ESS equipment, and testing of the systems began early the following year. In the spring of 1965, a small Number 1 ESS went into service in Succasunna, New Jersey, and 200 of the town's 4,300 subscribers enjoyed the benefits of new services such as call waiting, call forwarding, conference calls, and greater speed of connection. These new services, while simple for customers to use, represented a leap forward in the complexity and sophistication of switching equipment.[52]

Even with the new ESS system in production at Western Electric, Joel and his colleagues still had to sell the idea of electronic switching. During the first few years of ESS production he was frequently called upon to meet with representatives of the local Bell operating companies to encourage them to buy these machines. By 1967 each of the twenty-four Bell operating companies in the United States had installed one or more of the model Number 1 ESS.[53]

In the spring of 1962, the electronic switching laboratory moved to a new location in Holmdel, New Jersey. Instead of following his colleagues there, Joel accepted a new position at the old West Street lab in New York City. There he was put in charge of designing the many new ancillary devices needed to keep older switching technologies working in conjunction with newer telephone services. The thrust of this research was to find some way using electronics to allow Step-By-Step and Panel equipment, which represented a huge capital investment, to remain in service and fit into changing nationwide service plans. The coming of Direct Distance Dialing (including automatic identification of the calling line number), automatic pay telephone operation, Touch Tone service, and other features, made it necessary to continue to modify older switching equipment. Joel and his team devised machines that would store and manipulate incoming and outgoing information and control the older switching devices. One of Joel's missions during these years was to move the switchboard laboratory into the electronic switching era. The laboratory had already made some halting efforts in electronic switching, such as creating an electronic control module for one of the oldest of the electromechanical systems. However, Joel found that many of the existing staff were old-guard types with only electromechanical experience, a fact that caused many of the projects to continue to be based on electromechanical technology.

Joel soon changed this situation. The advent of Direct Distance Dialing technology automated about sixty percent of long-distance calls nationally. In early 1950s the company studied ways to automate the remaining forty percent. Calls requiring certain special procedures, such as collect calls or person-to-person calls, posed especially difficult problems for the designers of automatic switching. The studies commissioned by Bell Labs concluded that instead of attempting full automation, a more economical approach would be to develop a new kind of operator-attended switchboard, combined with Automatic Message Accounting.[54]

The Traffic Service Position (TSP), an earlier Bell Labs innovation, allowed an operator to manually enter data about operator-assisted, long-distance calls to Automatic Message Accounting devices. About twenty-one of these machines were installed beginning in 1963. They were integrated into the common control design of the No. 5 Crossbar system. In effect, the TSP was another Common Control device, this time in a semi-automatic form.

Joel faced the problem of adapting the TSP for use with the antiquated Step-By-Step system. Dick Jaeger, one of Joel's coworkers, suggested giving the engineers in the department, especially the newly hired younger men, some experience with electronics. The two proposed the idea of a stored-program control system to assist telephone operators. With it, AT&T could modify older switching systems to accept this new technology to expand the availability of the new national long-distance scheme. It provided automatic direct dialing while retaining the option of operator-assisted calls. Customers dialed the prefix "0" on special calls that required operator intervention or credit call number recording.

AT&T was soon persuaded of the merits of this idea and the system was renamed the Traffic Service Positions System (TSPS).[55] It became a major new component of the telephone network. TSPS was based loosely on the control unit in the Number 1 ESS, but used a new writable twistor memory. The first version of the TSPS, installed at Morristown, New Jersey, in 1969, served 3000 trunks with 320 operators. TSPS was adopted nationally, and by 1976 about half of all operator-assisted calls passed through TSPS offices. This new equipment resulted in a great reduction in the number of operators needed.[56]

TSPS presented new service opportunities as well. With it, for example, Bell could provide real-time toll-charge data to hotels for the purpose of billing patrons, and telephone credit-card numbers could be dialed and checked for fraud. It also allowed Bell customers to begin direct dialing to foreign countries from nonelectronic switching offices. Equipment was later added to TSPS offices to announce charges automatically to payphones using a 1949 invention of Joel's. These new services, instituted in the mid-1970s, proved highly profitable to the company by lowering operating costs for toll and long-distance calls. For this work, Joel and Jaeger received an Outstanding Patent award from the New Jersey Council for Research and Development.[57]

During these years, Joel's laboratory explored the possibilities of automatic directory assistance and intercept systems to provide information on number changes. For a while, the group experimented with an electromechanical switching device controlling a magnetic drum. The drum stored recorded messages about new and changed telephone numbers so that when someone called information to get the number of a new customer, or reached a changed number, the machine would automatically announce the new number. At the time this device proved not to be economically feasible, and it was not until later that such devices came to be installed in the network. Semiautomatic systems of this type, made by other manufacturers, appeared in 1965. It was not until Joel devised a way to use the line identification arrangements in electromechanical offices that it became possible to introduce a fully automatic intercept system in 1967. Like some other projects Joel worked on, automatic directory assistance technology experienced a long period of development and experimentation before being deemed sufficiently reliable and cost-effective for commercial service.[58]

ASPRAY: *Throughout your career at Bell Labs, do you feel that there was some sort of engineering style that characterized your personal work?*

JOEL: *I've always been motivated by wanting to invent, wanting to create something new. I always wanted to do something different. . . .*

ASPRAY: *Did you have a stock of tools that you used over and over?*

JOEL: *No, I had no tools. I had no one approach. . . . Ideas come to you in all different ways and different conditions. But in switching, it's usually combinations of things that trigger you off. I think it's important for people to know about history, because sometimes just thinking about how they did things way back, a century ago maybe [can give you ideas]. Doing that with modern technology may not be a bad idea. . . . I stress in my classes that people should always look back. It's not really re-inventing the wheel that you worry about, but people do tend to get certain things from relooking at old ideas with new technology. It can make a big difference. It may not just be the wheel, but getting a rubber tire on it and a few of those things that make it really worthwhile to have a wheel.*[59]

After several years of directing this switching laboratory, Joel tired of the routine of budgeting and management, desiring to use his technical rather than his managerial skills. He approached his boss, who reassigned him as a "director without portfolio." Thus, in 1967, Joel began circulating from laboratory to laboratory, making contributions to various projects when he felt he had something to add. After a short time, Joel was officially reclassified as a consultant within the company. Because he was free to exercise his creativity without the burden of administrative duties, he found his new status at Bell Labs invigorating. He appreciated the fact that Bell Laboratories allowed creative people to be creative.

One of the results of this freedom was his contribution to the early development of mobile telephones. After examining the cellular work in progress at Bell Labs, he realized that the engineers working on the new system had not yet designed suitable switching techniques. He offered his expertise in electronic switching, and this work brought him one of the basic patents on cellular communication in 1972. (See Figure 6.)

Bell Labs encouraged the participation of switching engineers in professional and teaching activities, and Joel took advantage of this opportunity. He promoted the idea of a special section of the American Institute of Electrical Engineers (AIEE) devoted to switching, and with a small group of experts he helped form the Switching Committee in the late 1940s. In the 1950s he became more involved with the AIEE looking for other ways to address the profession's failure to recognize properly the field of switching. He was active in initiating the International Switching Symposia and chaired the program committee of the 1972 meeting.

His involvement in education was no less extensive. In college, he noticed that switching was not part of the training in communication engineering as was telephone transmission. The latter had been elevated to an organized body of engineering theory, with textbooks written on the subject, while in the US switching was ignored. Based on the research he conducted for his master's thesis at MIT, Joel felt confident that the technology of switching could be taught at a college level. Before and during World War II he petitioned the top management at Bell Labs to let him try his ideas. When the war was over, it was recognized that the teaching of switching would be necessary as new engineers were added to the staff. As a result, a school Bell Labs established included a new switching curriculum in 1946, with Joel as one of the founding instructors. These educational opportunities, along with his collaboration on a switching textbook, allowed him to demonstrate the teaching of switching principles, rather than simply the description of switching systems.[60]

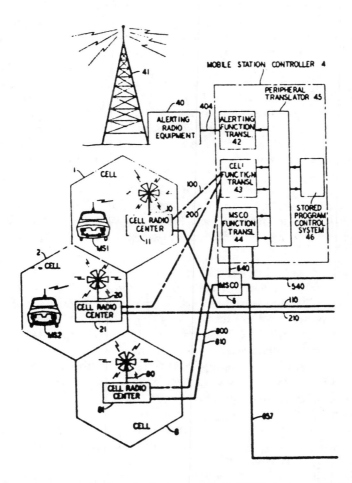

Figure 6. Illustration from the first page of Joel's 1972 patent on a cellular phone system.

Throughout his career, Joel continued to encourage young switching engineers to study the principles of switching, and he promoted the scientific
approach to the subject through publications. His courses and textbooks went
beyond the existing work on communications, or traffic theory, looking for the
generalized principles of switching system architectures. This approach has
been paralleled in the field of computer architecture. Although he left the project
soon after establishing it in order to pursue other interests, Bell Labs engineers
continued to offer these courses for many years and published a textbook based
on the course. Joel's efforts in promoting and developing this course are
recognized in the preface to this text.

The greatest engineering challenge has not been the specific inventions. . . .
The greatest challenge to me has been trying to fit together the state of the art of
switching and to make something of this. . . . We were able to show there are
certain principles—what switching is all about—that we can teach and explain
to succeeding generations. And as new switching techniques evolve, we can
continue to do that. I think we've put the framework down for doing that. Of
course, my big disappointment is we haven't been able to formalize it more. We
haven't been able to formalize it to the degree of putting some mathematics with
it that would allow one to evaluate and synthesize the various architectures and
so on. But at least I think people now understand the principles pretty well. When
I started out, there were no such things as identified principles. I couldn't go to
the vice president or even the engineer that had been working the longest on
switching systems at Bell Labs and say, "What kinds of switching systems are
there?" He'd tell me there was step-by-step and panel and manuals, but he
wouldn't be able to tell me the principles upon which they were based. I think
we've done an awful lot over the period of my career to do that. Today you can
classify the technologies and architectures that are used in switching.[61]

After several years of relatively independent engineering as consultant
at large within Bell, Joel's idyll came to an end. Periodically since its
founding, AT&T had been investigated by the federal government on
charges of unfair practices related to the company's monopoly position. In
early 1972 the Federal Communications Commission, the body formally
charged with regulating AT&T, began investigating anew. Because Joel
was an expert on the technologies of the telephone network and was not tied
down with other duties, company executives chose him to help with the
technical side of the company's defense.

In looking for instances of monopolistic practices, the FCC's gaze fell
upon the development of electronic switchgear, which AT&T dominated.
Joel was asked to provide information to defend against charges that AT&T
purposefully delayed the development of ESS in order to maintain its near
monopoly in the American market for central office equipment. This he did
until 1974, when the Justice Department initiated another investigation
and filed a new antitrust case against the company. Joel continued his

duties as an aid to the legal department, but ironically he was now defending the company against charges that such new technologies as ESS were introduced prematurely.[62]

In early 1983 the government concluded its presentation of the case, as AT&T reached a consent agreement with the Justice Department stipulating the divestiture of the Bell Operating Companies. This same year, Joel also reached his mandatory retirement age of sixty-five, so his somewhat frustrating research duties for the company came to a close.[63] In retirement, he still works with Bell Labs as a part-time consultant, and he also acted as a consultant to the then newly formed AT&T International. He also became a consultant to IBM, Continental Telephone, and some of the Bell operating companies.

In the course of his career, Joel published numerous technical articles and two historical monographs. In celebration of the centenary of the invention of the telephone in 1976, AT&T commissioned a series of books on the history of Bell Laboratories. Joel edited the volume on switching in the Bell System, which was published in 1983.[64] Just about that time, he also began work on another book, written with Robert Chapuis, on the history of electronic switching at Bell and internationally.

For his inventive and professional activities, Joel has been repeatedly honored by the engineering and scientific communities. His international recognition includes the Alexander Graham Bell Medal of the Institute of Electrical and Electronics Engineers (IEEE) and the Kyoto Prize Laureate in Advanced Technology by the Inamori Foundation. Local organizations honored him as well, including the New Jersey Congress of Inventors and the city of Genoa, Italy, which in 1984 awarded him a Columbian Medal and named him "Mr. Switching." His other honors include the IEEE Communications Society Achievement Award (1972), membership in the National Academy of Engineering (1981), the Franklin Institute's Stuart Ballantine Medal (1981), the International Telecommunication Union's Centenary Prize (1983), and AT&T's patent recognition award, the Charles E. Scribner Trophy (1992). In 1992 he received IEEE's highest award, the Medal of Honor.

Amos Joel, Electronic Switching, and the Innovation Process at Bell Laboratories

During the half century from the 1930s to early 1980s, Amos Joel devoted his professional career to the design of telephone switching systems. "Knowing a lot about one thing, namely switching," worked better for him than knowing "a little about a lot of things" like some other engineers.[65] But Joel was also constantly adapting himself to emerging new technologies, and he had the ability to see switching as part of a larger technological system. He continued to demonstrate this perspective even in retirement, as for example his recent patent on the emerging technology of photonic switching.

During the postwar period, Bell Labs carried forward its drive for greater speed, automation, and flexibility in switching and established a great many research projects in electronics. Although electronic switching appeared to emerge suddenly in the 1970s and 1980s, this revolution actually began quietly in the 1950s. Joel was instrumental in producing the series of innovations that proved the concept of electronic switching, even though they sometimes had little immediate impact on the public network. The accumulated knowledge gleaned from such experiments as the MDAS and the Remote Line Concentrator contributed directly to the full-scale Number 1 ESS. From an engineering perspective, electronic telephone switching also represented a number of technical breakthroughs:

> The [Morris] system proved that large numbers of new electronic devices could be assembled and operated reliably. It also demonstrated the feasibility and forecast public acceptance of some of the new services that the low cost memory and the flexibility of stored program control made practicable. These services included call transfer, add-on conference, abbreviated dialing, code calling of extension telephones, series completion of calls to nonconsecutive line numbers.[66]

Joel's career illustrates several important features of Bell Labs. AT&T's special position, when it had a virtual monopoly, helped to reinforce the tendency toward long-term, company-wide planning for technological change. This high degree of organization and long-term planning did not preclude competition among different projects at Bell Labs. More than once, electronic equipment devised by Joel vied for sponsorship with electromechanical and other electronic equipment performing the same functions. Bell Labs had the financial wherewithal to investigate, often simultaneously, several different answers to technical problems in a way that probably would not have been possible in a less secure, free-market situation. Even after some of Joel's innovations successfully passed from design to field test, the company had the ability to reject—and sometimes did reject—them as being "impractical" or "not economical at present." AT&T, though it felt compelled to respond to the public's demand for improved service at a lower cost, set its own pace of technological innovation and its own high standards of reliability for implementation.[67]

The development of electronic switching tells much about the politics of innovation at Bell Laboratories. Those like Joel who believed in the promise of electronics faced numerous obstacles above and beyond technical difficulties, including resistance from the supporters of competing technologies both inside and outside the Labs. Electromechanical switching represented the accumulated knowledge of decades of invention and innovation, and had repeatedly proven its viability. It was not surprising that many Bell Labs engineers, AT&T managers, and telephone customers resisted the relatively unproven electronic technology. On the other hand, Joel's experience shows how the sponsorship of laboratory directors and managers could ease the way for a controversial technological change. In many ways, electronic switching was the vision of upper managers and engineers at the department head or director level. The process of creating the first ESS involved communicating this vision down the hierarchical ladder. Joel's involvement in the engineering, planning, and selling of electronics at Bell demonstrates this process at every level.[68]

My family life has been, I think, a very important contributor to my ability to do the things that I've done. . . . it's been a very tranquil family life. . . . My wife and I are celebrating our 50th anniversary this year, and we've had a very good life together. . . . Our three children have all been a joy.

As far as my hobbies are concerned, I've always been interested in music. . . . I did have a period when I was rather ill . . . and the doctor found out that I was a workaholic. He said, "You've got to have hobbies." So I decided I would like to play the organ. . . . my wife went out and rented an organ, and once we had an organ in the house, we never let it go. Kept buying newer and bigger and better. Right now I just got a new gadget for my birthday that's coming up. I just enjoy playing the organ. For the last 15 years or so, every night after supper I sit down and play the organ for at least a half hour. I get a lot of relaxation and pleasure out of that. I wouldn't say I play well, but it's a good hobby. Of course, I get involved in the technology of it, and so I've got all kinds of gadgets to go with it. I don't just play straight, I play these other things as well—unfortunately, my wife doesn't like this.

Through the last 20 years at Bell Labs, I always had access to computers, and have used them a great deal in forming databases and doing research and writing papers. . . . I've spent a lot of time keeping up to date with computers and enjoy that as a hobby, too.

It was hard to get me to take a vacation for many years. Since I've been retired, I have enjoyed taking vacations—particularly cruising. I like to take a boat. I don't take my work with me, and so I enjoy it much more that way. My wife says it's because there are no telephones to answer.[69]

---

[1] For this chapter, the principal source of personal information about Amos Joel is an extended oral history interview of Joel conducted by William Aspray 4 and 18 February 1992 (referred to here as Interview 1992). An edited transcript is held at the IEEE Center for the History of Electrical

Engineering. (The passage quoted here is from page 3.) The author wishes to thank the following people who contributed to this chapter either by providing factual information or by offering critique and commentary: Amos Joel, William Aspray, Andrew Goldstein, Frederik Nebeker, and Steven Vallas.

[2] Interview 1992, p. 6.

[3] Interview 1992, pp. 8–9, slightly edited.

[4] During these years, many top engineering schools modernized their electrical engineering curriculums in a similar way. See Frederick Terman, "A brief history of electrical engineering education," *Proceedings of the IEEE*, vol. 64, 1976, pp. 1400–1401.

[5] John Horgan, "Claude E. Shannon," *IEEE Spectrum*, vol. 29, no. 4, 1992, p. 72.

[6] Interview 1992, p. 25.

[7] Ibid.

[8] The formation of the Bell Laboratories is discussed in Leonard Reich, *The Making of American Industrial Research: Science and Business at GE and Bell, 1876–1926* (Cambridge: Cambridge University Press, 1985), pp. 153–154; R. F. Rey, ed., *Engineering and Operations in the Bell System* (Murray Hill NJ: AT&T Bell Laboratories, 1983), p. 24. It should be noted that Bell's employment figures for technical personnel do not represent the true number of persons employed. Technicians and those without engineering degrees were counted only half, while degreed engineers and other "full" members of the technical staff were each counted once. [M. D. Fagan, ed., *A History of Engineering and Science in the Bell System: National Service in War and Peace 1925–1975* (Bell Telephone Laboratories, Inc.; 1978), p. 11.]

[9] By the late 1930s, the practice of training engineers by moving them from department to department was well-established at many large American manufacturing companies. AT&T and other large manufacturers also maintained strong ties to major engineering institutions and often provided the faculty for specialized engineering courses. See Ronald R. Kline, "Origins of the issues," *IEEE Spectrum*, vol. 21, no. 11, 1984, pp. 38–43; and Kline, "The General Electric Professorship at Union College, 1903–1941," *IEEE Transactions on Education*, vol. 31, 1988, pp. 141–147.

[10] Fagan, *A History of Engineering and Science*, pp. 9–14.

[11] Interview 1992, p. 46.

[12] This machine was based on the World War I–era work of another Bell Labs researcher, G. S. Vernam, and undertaken as a collaboration between AT&T and the Teletype Corporation. Teletype machines of the day transmitted a letter or number by converting it into a five-bit binary code. Vernam's plan called for both transmitting and receiving Teletype machines to have matching copies of a special punched tape, which supplied a random, nonrepeating source of additional code pulses. The machine added the two pulses to produce the coded message and transmitted it just as with an uncoded Teletype message. At the receiving end, the machine could synchronize itself to the incoming message, combine the incoming pulses with the pulses supplied by the matching punched tape, and print the deciphered message. See Fagan, *A History of Engineering and Science*, pp. 244–246; and E. F. Watson, "Fundamentals of teletypewriters used in the

Bell System," *Bell System Technical Journal*, vol. 17, 1938, pp. 620–639. AT&T purchased the Teletype Corporation in 1930, and it is currently a subsidiary of Western Electric [Rey, *Engineering and Operations in the Bell System*, p. 5].

[13] Fagan, *A History of Engineering and Science*, pp. 298–306; Homer Dudley, "The carrier nature of speech," *Bell System Technical Journal*, vol. 19, 1940, pp. 495–515.

[14] Fagan, *A History of Engineering and Science*, p. 308. Interestingly, the patents filed between 1941 and 1945 on this coded transmission system were kept secret, some for over three decades. See US Patent 3,967,067 (29 June 1976), Ralph K. Potter, "Secret telephony"; and US Patent 3,985,958 (12 October 1976), Homer W. Dudley, "Secret telephony"; Rather similar work was conducted by the US Army Signal Security Agency on machines used to analyze and decipher telegraphic messages sent by German Enigma machines. See David J. Crawford and Philip E. Fox, "The Autoscritcher and the Superscritcher: Aids to cryptanalysis of the German Enigma Cipher Machine, 1944–1946," *IEEE Annals of the History of Computing*, vol. 14, 1992, pp. 9–22. Joel's work on secrecy systems resulted in several other patents which were kept secret for over 25 years. One of these included the basic idea of creating a long key code by the use of prime number generators [Joel, private communication, 10 February 1993].

[15] See Fagan, *A History of Engineering and Science*, p. 170; and E. G. Andrews, "Telephone switching and the early Bell Laboratories computers," *Bell System Technical Journal*, vol. 42 , 1962, pp. 341–354.

[16] For more information on this so-called CADET (Can't Add, Doesn't Even Try) architecture or on the specifications of the various Bell Labs relay computers, see chapter 6 of Michael R. Williams, *A History of Computing Technology* (Englewood Cliffs NJ: Prentice-Hall, 1985). These developments should not be confused with other wartime work at Bell Labs on electric analog calculators used in the field for gun-aiming.

[17] William Keisler, et al., *The Design of Switching Circuits*,(Princeton NJ: D. Von Nostrand, 1951), pp. 194–199.

[18] F. A. Korn, "The Number 5 Crossbar System," *Transactions of the AIEE*, vol. 69, 1950, pp. 244–254; and Keister, *The Design of Switching Circuits*, pp. 446–449.

[19] F. J. Scudder and J. N. Reynolds, "Crossbar dial telephone switching systems," *Bell System Technical Journal*, vol. 18, 1939, pp. 76–118.

[20] A. B. Clark and H. S. Osborne, "Automatic switching for nationwide telephone service," *Bell System Technical Journal*, vol. 31, 1952, pp. 823–831; W. H. Nunn, "National numbering plan," *Bell System Technical Journal*, vol. 31, 1952, pp. 851–859; F. F. Shipley, "Automatic toll switching systems," *Bell System Technical Journal*, vol. 31, 1952, pp. 861–862. The national numbering plan also signaled the abandonment of the old mnemonic devices once used by AT&T to make it easier for Americans to remember telephone numbers. Whereas telephone numbers were previously identified using a combination of letters and digits (e.g., PEnnsylvania 6-5000), in combinations of up to seven digits, eventually (in the period between 1958 and 1977) they were uniformly composed solely of a 3-digit area code plus seven digits [Richard Brodsky, "Your evolving phone number," *American Heritage of Invention and Technology*, vol. 6, 1991, p. 64].

[21] In taking this project, Joel inherited the results of several years of research on automatic accounting. Bell researchers had been tinkering with such devices since the late 1930s and had developed a device that could automatically produce cards printed with information about individual calls, such as the originating number, the number called, and the duration of the call. In 1944 AT&T installed this experimental automatic ticketing machine in the Los Angeles network. See A. E. Joel, et al., *A History of Engineering and Science in the Bell System: Switching Technology 1925–1975* (Bell Telephone Laboratories; 1982), pp. 132–135; and Joel, "Tracing time backward in AMA," *Bell Laboratories Record*, vol. 30, 1952, pp. 422–429.

[22] Joel, *A History of Engineering and Science*, pp. 142–143; G. V. King, "Centralized automatic message accounting system," *Bell System Technical Journal*, vol. 33, 1954, pp. 1331–1342. Joel also worked on a number of other devices in the immediate postwar years. He received a patent for improvements to coin-operated calling for a system that would automatically register the value of coins inserted into a pay telephone. Previously, the operator had to listen to the sound of the coins as they were inserted. However, this idea was left on the shelf until the 1970s, when it was revived and modified for use with the new long-distance pay station direct dialing service. See page 259 of this chapter and Interview 1992.

[23] Even the earliest AT&T electronic switchgear gave an indication of the speed increases possible through the use of electronics. Vacuum tube diodes were 10 times faster than the best electromechanical relays, while transistors were 1000 to 10,000 times faster. See Joel, "An experimental switching system using new electronic techniques," *Bell System Technical Journal*, vol. 37, 1958, pp. 1098–1099.

[24] Joel, *A History of Engineering and Science*, pp. 43–44, 200–201.

[25] One Bell engineer in 1965 stated succinctly AT&T's public justifications for adopting electronic switching, saying that postwar work in this area was motivated by the company's desire to "improve service, reduce costs, and provide greater flexibility while retaining high reliability." See W. H. C. Higgins, "A survey of Bell System progress in electronic switching," *Bell System Technical Journal*, vol. 44, 1965, p. 938. The electromechanical Private Branch Exchange was also in use at this time. This class of switch, still in use, is put into large buildings and offices to distribute incoming and outgoing calls between a multitude of different telephones and a smaller number of wires to the central office. Its value is that it removes the necessity to supply a separate pair of wires from the central office to each telephone, and therefore the PBX saves money. When a PBX is installed in an office building, there is the assumption that not all the telephones in the building will be in use simultaneously. A telephone call to any telephone in the building is transmitted from the central switching office through a common pair of wires (the number of physical lines to the building is based on the number of telephones in the building and the expected traffic) and then distributed to the proper telephone by the PBX. Similarly, when the PBX detects an outgoing call, it automatically makes a connection between one of the large number of telephones in the building and one of the small number of lines to the central office.

[26] Joel, *A History of Engineering and Science*, pp. 214–215. Joel's 1956 technical paper on the Line Concentrator explained the objections that AT&T had regarding decentralized switching; the largest concern was the difficulty of maintenance and the cost of controlling switchgear at a distance

[Joel, "Experimental remote controlled line concentrator," *Bell System Technical Journal*, vol. 35, 1956, p. 250].

[27] Joel, *A History of Engineering and Science*, p. 215.

[28] Ironically, the line concentrator idea was so successful that another laboratory built an electromechanical version of it. This device, called the Universal Concentrator, used a miniature, modified crossbar switching system. It was of a scale similar to Joel's concentrator, serving 50 to 100 individual lines over ten trunk lines to the central office. Unlike the Remote Line Concentrator, the Universal Concentrator was widely used. In all, about 4,300 of these devices were installed around the country, beginning in 1961. Most were in use only temporarily, allowing local exchanges to delay the installation of additional telephone lines. In the end, maintenance costs proved to be too high to justify permanent installation.

Still, the Universal Concentrator demonstrates the ingenuity that Bell's electromechanical switching designers displayed in reaction to an unfamiliar new technology, electronics. According to Joel, many of them felt that their jobs were threatened by electronics, and lobbied against it within the organization. Joel encountered this resistance repeatedly, and it contributed to the problems AT&T encountered as it tried to convert the network to electronic switching. [Joel, *A History of Engineering and Science*, p. 384; "Telephone interview with Amos E. Joel, Jr., by David L. Morton, Jr.," 17 October 1992 (hereafter Morton, "Joel Interview")].

[29] Interview 1992, pp. 88–91, slightly edited.

[30] Joel, *A History of Engineering and Science*, pp. 222–223; Robert J. Chapuis and Amos E. Joel, Jr., *100 Years of Telephone Switching*, vol. 2, *Electronics, Computers and Telephone Switching* (Amsterdam: North-Holland Publishing Company, 1990), p. 44.

[31] Chapuis and Joel, *100 Years*, p. 191; Joel, *A History of Engineering and Science*, p. 24. As in the case of the Remote Line Concentrator, an electromechanical version of the MDAS emerged. This was the Auxiliary Sender, a machine that enabled older electromechanical systems to handle the extra digits used in Direct Distance Dialing. Starting in 1957, Auxiliary Senders were added to many such systems. These were in place for about ten years before finally being superseded in 1967 by an entirely new electronic device.

[32] Interview 1992, p. 73.

[33] Charles Weiner, "How the transistor emerged," *IEEE Spectrum* vol. 30, no. 1, 1973, p. 30.

[34] Joel's first opportunity to join a browser group was in late 1941, when he was invited to join a group interested in new switching technologies. Before he was able to begin this work the US entered World War II and Joel was reassigned to war work. See Interview 1992.

[35] Chapuis and Joel, *100 Years*, p. 48.

[36] Bell Labs and Western Electric worked together on many projects. See McKinsey and Co., Inc., *A Study of Western Electric's Performance* (New York: American Telephone and Telegraph, 1969), p. 72.

[37] Interview 1992, p. 89.

[38] Joel, *A History of Engineering and Science*, p. 225.

[39] Ibid., pp. 225, 229; Joel, "Experimental electronic switching system,"

Bell Laboratories Record, vol. 36, 1958, pp. 359–363.

[40] C. W. Hoover, Jr., R. E. Staehler, and R. W. Ketchledge, "Fundamental concepts in the design of the Flying Spot Store," *Bell System Technical Journal*, vol. 37, 1958, pp. 1161–1194; Joel, *A History of Engineering and Science*, pp. 239–240.

[41] In the ESS, the function allowing multiple inputs was called "scheduling." See Chapuis and Joel, *100 Years*, pp. 113–114. One of the few multiple-input, real time computers of the day was the Whirlwind, developed in the late 1940s at MIT. This machine, infamous in its day for its cost and complexity, was later incorporated into the SAGE radar air defense system, developed in the early 1950s by the US Air Force. In its experimental form, the SAGE system accepted data from several radar sets, interpreted it, and formed a composite radar image on a cathode-ray tube display. See George E. Valley, Jr., "How the SAGE development began," *Annals of the History of Computing*, vol. 7, 1985, pp. 196–226.

[42] Joel, "Experimental Switching System," pp. 1091–1124.

[43] Ibid., p. 240.

[44] R. R. Everett, et al., "SAGE—a data processing system for air defense," *Annals of the History of Computing*, vol. 5, 1983, pp. 330–339.

[45] Joel, *A History of Engineering and Science*, pp. 14–15, 238, 240.

[46] Interview 1992, p. 96.

[47] Interview 1992, p. 97.

[48] During the Morris trials, machine-related "errors" affected about 7.5 percent of all calls at the time of start-up, dropping to less than 1 percent after 7 months. The customers whose telephones were connected to the Morris switch were given the opportunity to experiment with abbreviated dialing (a 2-digit speed dialing system for frequently called numbers), code dialing (a household intercom feature using the telephone), and call forwarding. On average, 15 percent of all calls were placed using the abbreviated dialing, and in some households that figure was as high as 50 percent. See Higgins, "Survey," pp. 942–943.

[49] Another ferromagnetic memory device, called the ferrite sheet, was used for a time before the magnetic core memories were developed. By the mid-1950s, most computer builders had begun to use magnetic core memories. Western Electric did not want to use cores because they could not make them at a price competitive with the rest of the industry (which was having them assembled in Japan). Instead, Bell Labs opted to develop the ferrite sheet, which functioned in a similar fashion but was designed to be made economically by Western Electric. See Chapuis and Joel, *100 Years*, p. 155.

[50] The Ferreed was, according to Joel, "a major bridge between electronics and electromechanical contacts," in that they "responded to short pulses, they operated rapidly (less than one millisecond), they magnetically latched, [and] they required no holding power . . . " [Chapuis and Joel, *100 Years*, p. 154]. The Ferreed had contacts sealed in a small glass tube, with an electromagnet around the tube. A pulse of current in the electromagnet closed the contacts. Initially there was an external permanent magnet; later the reeds stayed "latched" due to their own magnetic attraction. When a subsequent path conflicted with operated contacts, a pulse of opposite polarity in the electromagnet reopened the contacts. See Chapuis and Joel, *100 Years*, pp. 54–55; A. Feiner, C. A. Lovell, T. N. Lowry, and P. G. Ridinger,

"The Ferreed—a new switching device," *Bell System Technical Journal*, vol. 39, 1960, pp. 1–30.

[51] Joel, "Electronics in telephone switching systems," *Bell System Technical Journal*, vol. 35, 1956, p. 994. The computing logic used in the ESS was adapted from an earlier project, the Electronic PBX (EPBX). These circuits required more expensive silicon diodes and transistors, but were faster. See Chapuis and Joel, *100 Years*, pp. 248–249.

[52] The *Bell System Technical Journal* for September 1964, comprising over 700 pages, was devoted entirely to the Number 1 ESS. Scanning the 19 articles on the various ESS subsystems reveals that dozens of engineers made technical contributions. Many other contributors are not named. The special issue of the *Journal* demonstrates the extent to which the ESS was a corporate effort. It is interesting to note that by the time the issue appeared, Joel had been moved to other projects and his name did not appear in connection with the system he had done so much to bring about.

[53] Chapuis and Joel, *100 Years*, p. 158; Morton, "Joel Interview."

[54] Joel, *A History of Engineering and Science*, pp. 307–308.

[55] "Traffic" refers to incoming and outgoing calls, and "position" refers to where the human operators are seated.

[56] In 1970, when the rate of installation of electronic switchgear was beginning to increase, Joel wrote that "eliminating the human element in these large networks has not only accelerated their growth, but has in many cases resulted in their being engineered on a non-delay basis." See Joel, "Twenty-five years of switching system innovation," *Proceedings of the National Electronics Conference*, 1970, p. 883.

[57] Joel, *A History of Engineering and Science*, pp. 311–315.

[58] Ibid., pp. 322–324.

[59] Interview 1992, pp. 149–150, edited.

[60] The first AT&T switching textbook to which Joel contributed has been cited previously as William Keister, et al., *The Design of Switching Circuits*. Joel also contributed to the later "Orange Books," including *Switching Circuits and Systems* (New York: American Telephone and Telegraph Company, 1961).

[61] Interview 1992, pp. 175–176, edited.

[62] United States District Court for the District of Columbia, Civil Action 74–1698, *United States vs. AT&T et al.*, Plaintiff's third statement of contentions and proof, reprinted in Christopher H. Sterling, et al., eds., *Decision to Divest: Major Documents in U.S. vs. AT&T, 1974–1984* (Washington, D.C.: Communications Press, 1986), pp. 1578–1724.

[63] Joel felt that his creative role in the company was nearly overwhelmed by the duties of his last years at Bell Labs, describing those duties as "onerous." His friends often commented, "Think how many additional inventions he might have made if he had not been given these assignments." [Joel, private communication, 10 February 1993.]

[64] A. E. Joel, et al., *A History of Engineering and Science in the Bell System: Switching Technology 1925–1975* (Bell Telephone Laboratories; 1982).

[65] Interview 1992, pp. 176–177.

[66] Chapuis and Joel, *100 Years*, p. 53.

[67] Many Bell Labs projects were expensive and involved long-term commitments to research before any economic return was realized by the company. One example, the development of the transistor, is discussed in Ernest Braun and Stuart McDonald, *Revolution in Miniature* (Cambridge: Cambridge University Press, 1978), p. 40.

[68] The same sort of process seems to have been at work in the development of semiconductors at Bell. Historian Charles Weiner writes of Bell Laboratories' executive vice president Mervin J. Kelly's crucial role in promoting semiconductor research within the organization. See Weiner, "How the transistor emerged."

[69] Interview 1992, pp. 178–181, edited.